Pricing and Forecasting Carbon Markets

Bangzhu Zhu · Julien Chevallier

Pricing and Forecasting Carbon Markets

Models and Empirical Analyses

 Springer

Bangzhu Zhu
School of Management
Jinan University
Guangzhou
China

Julien Chevallier
IPAG Lab
IPAG Business School
Paris
France

and

University Paris 8 (LED) UFR AES
  Economie Gestion
Saint-Denis Cedex
France

ISBN 978-3-319-86209-5   ISBN 978-3-319-57618-3   (eBook)
DOI 10.1007/978-3-319-57618-3

Printed on acid-free paper

This Springer imprint is published by Springer Nature
The registered company is Springer International Publishing AG
The registered company address is: Gewerbestrasse 11, 6330 Cham, Switzerland

Foreword

As a policy tool of the trading mechanism, carbon market is a great institutional innovation for coping with global climate change. Due to its multiple advantages of saving cost, protecting environment, and political feasibility, more and more countries including China have applied carbon market for CO_2 emissions reduction.

During the recent years, the price of global carbon market, represented by the European Union Emissions Trading System, exhibits a great fluctuation. This significantly affects the performance for CO_2 emission reduction and results in a great loss of China's carbon assets. Accurately understanding the pricing mechanism of carbon market is essential to build a national carbon market for China, in which there are a series of issues of management science and energy economics. Therefore, pricing and forecasting carbon market, and the related issues have been aroused both concerns of researchers and practitioners.

Unlike the conventional financial markets, carbon market, as a policy-based artificial market, is influenced by both the market mechanisms and the external heterogeneous environments. Especially, various factors are subject to changeful interpenetration and complex nonlinear dynamic relationships, which leads to the complexity of the pricing behavior for carbon market. Prof. Bangzhu Zhu and Prof. Julien Chevallier have explored the related issues of pricing and forecasting carbon market from the perspectives of theoretical models and empirical analyses in this book. Thus, this book is of significance with innovation, advancement, operability, and practicality.

It is expected to be a preferable book integrating the features of analytical system, and a certain depth and far sight. The publication of this book is beneficial for further scientifically understanding the pricing mechanism of carbon market. Moreover, it lays a foundation for, and enriches the knowledge of, dealing with the climate change for China and the construction of her own national carbon market. In addition, it will actively contribute to the energy saving and CO_2 emission reduction promoted by the carbon market.

December 2016   Prof. Yi-Ming Wei
Director, Center for Energy and Environmental
Policy Research, Beijing Institute of Technology
Beijing, P.R. China

Contents

Acronyms

ADF	Augmented Dicky-Fuller
AI	Artificial Intelligence
AIC	Akaike Information Criterion
ANN	Artificial Neural Network
ARIMA	Autoregressive Integrated Moving Average Model
ARMA	AutoRegressive Moving Average
BDS	Brock-Decher-Scheikman
BIC	Bayesian Information Criteria
BP	Back Propagation
CDM	Clean Development Mechanism
CER	Certified Emission Reduction
CO_2	Carbon Dioxide
CRSP	Centre for Research on Securities Prices
DCA	Directional Change Accuracy
DM	Diebold-Mariano
ECX	European Climate Exchange
EELM	Extended Extreme Learning Machine
EEMD	Ensemble Empirical Mode Decomposition
EEX	European Energy Exchange
EMD	Empirical Mode Decomposition
EPPA	Emissions Prediction and Policy Analysis
ES	Event Study
EU	European Union
EU ETS	European Union Emissions Trading System
EUA	European Union Allowance
FAVAR	Factor-Augmented Vector Autoregression
GARCH	Generalized AutoRegressive Conditional Heteroskedasticity Model
GHG	Greenhouse Gas
HAR-RV	Heterogeneous Autoregressive Model for Realized Volatility
ICE	Intercontinental Exchange

ICSS	Iterative Cumulative Sums of Squares
IMF	Intrinsic Mode Functions
LSSVM	Least Square Support Vector Machine
LSSVR	Least Square Support Vector Regression
MAE	Mean Absolute Error
MAPE	Mean Absolute Percentage Error
MLRM	Multiple Linear Regression Model
MR	Multivariate Regression
MS	Markov Switching
MSPE	Mean Square Prediction Error
NAP	National Allocation Plans
NASA	National Aeronautics and Space Administration
OLS	Ordinary Least Squares
OTC	Over-the-counter
PSO	Particle Swarm Optimization
PSR	Phase Space Reconstruction
PT	Pesaran-Timmermann
RBF	Radial Basis Kernel Function
RMSE	Root Mean Square Error
RT	Ratio Test
S&P	Standard Poor
SAR	Standard Abnormal Return
SPA	Superior Predictive Ability
SRM	Structural Risk Minimization
SSR	Sum of Squared Residuals
SVM	Support Vector Machines
SVR	Support Vector Regression
TD	Transaction Day
TGARCH	Threshold Generalized AutoRegressive Conditional Heteroskedasticity Model
UD	Uniform Design
UNFCCC	United Nations Framework Convention on Climate Change
VAR	Vector Autoregression
VIF	Variance Inflation Factor

List of Figures

List of Tables

Introduction

Global climate change has been one of the greatest challenges in the twenty-first century. Carbon market, represented by the European Union Emissions Trading System (EU ETS), is a cost-effective measure for tackling the climate change. Furthermore, pricing and forecasting carbon market has been one of the research focuses in the fields of energy and climate change.

In this book, the multidiscipline approaches of econometrics, statistics, finance, and artificial intelligence are used for pricing and forecasting carbon market with the following issues.

Chapter 1 provides an accessible introduction to the importance, literature review, and architecture of this book.

Chapter 2 explores the drivers of carbon price using the structure breakpoint test, cointegration techniques, and ridge regression.

Chapter 3 examines the structural changes of European carbon futures price using the iterative cumulative sums of squares (ICSS) algorithm and event study models. Special thanks to Shujiao Ma and Yi-Ming Wei in assisting the writing of Chap. 3.

Chapter 4 examines the drivers of European carbon futures price using ensemble empirical mode decomposition (EEMD) from a perspective of multiscale analysis. Special thanks to Ping Wang, Dong Han, and Ying-Ming Wei in providing research assistance for Chap. 4.

Chapter 5 investigates the European carbon futures price dynamics by applying the Zipf analysis. Special thanks to Shujiao Ma, Lili Yuan and Ying-Ming Wei for collaborating research on Chap. 5.

Chapter 6 proposes a hybrid ARIMA and least squares support vector machine (LSSVM) model for carbon price forecasting. Special thanks to Lili Yuan and Ying-Ming Wei for supporting writing of Chap. 6.

Chapter 7 develops a parameters simultaneous optimization of phase space reconstruction (PSR) and LSSVM with uniform design for carbon price forecasting so as to obtain high forecasting accuracy and high modeling efficiency. Special thanks to Xuetao Shi, Dong Han, Ping Wang and Ying-Ming Wei for actively counseling to the writing of Chap. 7.

Chapter 8 proposes a multiscale prediction model hybridizing empirical mode decomposition (EMD), particle swarm optimization (PSO), and LSSVM to predict carbon price. Special thanks to Dong Han and Yi-Ming Wei for helping writing on Chap. 8.

Chapter 9 develops an adaptive multiscale ensemble learning paradigm incorporating EEMD, PSO, and LSSVM with kernel function prototype to forecast nonstationary and nonlinear carbon price. Special thanks to Xuetao Shi, Ping Wang, Dong Han and Ying-Ming Wei for commenting the research writing of Chap. 9.

The writing of this book is conducted by Prof. Bangzhu Zhu and Prof. Julien Chevallier. This book is also the pearl of our research teams' joint efforts. Ying-Ming Wei, Ping Wang, Hua Liao, Dong Han, Lili Yuan, Xuetao Shi, Shujiao Ma, Xueping Tao, Sidong Liu, Kefan Wang, Minxing Jiang, and Runzhi Pang participated in the related research, discussion, and proofreading of certain chapters. Our most sincere thanks should be given to each member of our research teams.

We are most grateful to numerous professors including Yi-Ming Wei, Ziyou Gao, Jingyuan Yu, Zhaohan Sheng, Xiaotian Chen, Yijun Li, Shouyang Wang, Haijun Huang, Liexun Yang, Zuoyi Liu, Ruoyun Li, Gang Wu, Zhongfei Li, Weiguo Zhang, Xiangzheng Deng, Yong Geng, Lean Yu, Ying Fan, Jianping Li, Fan Wang, Lixin Tian, Dequn Zhou, Zhaohua Wang, and Peng Zhou for their helpful instruction and supports on our research into energy and carbon markets, as well as energy economics and climate policy since 2009.

Our most sincere thanks will give to Prof. Jun Hu and Xianzhong Song who serve as the President, and Vice President of Jinan University, China, respectively, as well as other colleagues including Jie Zhang, Yaohui Zhang, Haiying Wei, Yuyin Yi, Guoqing Wang, Bing Wang, Xia Wei, Hongtao Shen, and Jingyan Fu for their supports.

We should express my gratitude to the National Natural Science Foundation of China (71473180, 71201010 and 71303174), National Philosophy and Social Science Foundation of China (14AZD068, 15ZDA054), Natural Science Foundation for Distinguished Young Talents of Guangdong (2014A030306031), Guangdong Young Zhujiang Scholar (Yue Jiaoshi [2016]95), Department of Education of Guangdong ([2013]246, [2014]145), Guangdong key base of humanities and social science: Enterprise Development Research Institute and Institute of Resource, Environment and Sustainable Development Research, and Guangzhou key base of humanities and social science: Centre for Low Carbon Economic Research for funding supports.

We should also acknowledge all the authors of the cited literatures. There may be some shortfalls in this book due to the limited knowledge of the authors. If there are any opinions, please do not hesitate to let us know via e-mails: wpzbz@126.com (Bangzhu Zhu) and/or jpchevallier@gmail.com (Julien Chevallier).

February, 2017   Prof. Bangzhu Zhu
                                                             Prof. Julien Chevallier

Chapter 1
New Perspectives on the Econometrics of Carbon Markets

Abstract This chapter provides an accessible introduction to this book. First, we detail the importance of pricing and forecasting carbon market. Second, we review the pricing and forecasting carbon market from the perspectives of carbon price drivers, single scale forecasting and multiscale forecasting. Third, we provide the architecture of this book, and summarize the chapters.

1.1 Significance of Pricing and Forecasting Carbon Market

As the Kyoto Protocol took into effect in 2005, greenhouse gas emission permit has been a scarce resource which is endowed with a commodity attribute. Under such circumstance, carbon market was come into being in the fields of dealing with the global climate change. Global carbon market, represented by the European Union Emissions Trading System (EU ETS) has witnessed a rapid development: its turnover increased to 176 billion USD in 2011 from 10 billion in 2005, with an annual growth rate of 60%, so that it is expected to be one of the biggest and most active trading markets in the world.

In this new background, carbon price had a violent fluctuation in. Carbon price had been increasing to 35 Euro/CO_2 equivalent in April 2006 compared to 16 Euro/CO_2 equivalent in its initial stage in April 2005, which was the highest in a new historic record. Since May 2006, the leakage of verified data led to a sharp decrease of carbon price: carbon price dropped to 10 Euro/CO_2 equivalent. Carbon price gradually rose due to the European Union (EU) stricter CO_2 emission reduction policy in January 2007, and reached a new highest point, 35 Euro/CO_2 equivalent, in May 2008. However, carbon price showed a continuous drop due to the global financial crisis since July 2008. As the financial crisis was eased since February 2009, carbon price rebounded after dropped to the lowest point. However, it slightly decreased owing to the European debt crisis. Nowadays, carbon price is still below 10 Euro/CO_2 equivalent, which hits the new history record at lowest point.

© Springer International Publishing AG 2017
B. Zhu and J. Chevallier, *Pricing and Forecasting Carbon Markets*,
DOI 10.1007/978-3-319-57618-3_1

1

China, as the biggest provider for clean development mechanism (CDM) in the world, has provided lots of certified emission reductions (CERs) for global carbon market. However, carbon price shows a violent fluctuation with complexity, which giving rise to the serious loss of China's carbon assets: China's loss in carbon assets due to its price difference reached to 3.3 billion Euros in 2008 (Yang 2010). The reasons can be two bold: one, China has no pricing rights for the lack of her own carbon market. The other, there are few effective pricing and forecasting carbon market. Inversely, the various losses induced by price difference would be reduced so far as to be avoided at maximum with good precautions.

Although pricing and forecasting carbon market is very important and has attracted more and more attentions, there is no much obvious progress being made. On the whole, the existing methods used for pricing and forecasting carbon market can be roughly classified into two groups: econometric models and artificial intelligence approaches. However, these approaches cannot perform well on the real data of carbon price due to following reasons: as carbon market is a typical complex system of social economy, its price has uncertainty, nonlinearity, mutation, and instability (Chevallier 2011b) due to the interactions among multiple factors and their external heterogeneous environments, as well as their influences. This makes that these methods are unlikely to achieve satisfactory performance on the pricing and forecasting carbon market. Under such circumstance, it is worthy of performing a multiscale forecasting analysis. The multiscale forecasting analysis can decompose the complex carbon price time series into simple modes, with a simpler, more stable and more regular structures than the original carbon price time series into simple modes, which is more easily to be explored and forecasted (Zhu 2012).

Under the background of nonstationary and nonlinear, the multiscale forecasting analysis can significantly improve the accuracy of pricing and forecasting carbon market, which is not only beneficial for avoiding or reducing unnecessary losses in the CDM project for China, but also helpful for the construction of China's national carbon market. Therefore, it is of academic and practical significances for this book to use the multidiscipline approaches of econometrics, statistics, finance, and artificial intelligence for pricing and forecasting carbon market.

1.2 Review of Pricing and Forecasting Carbon Market

1.2.1 Carbon Price Drivers

Energy prices, external heterogeneous environments, temperature conditions, and economic activity are the main drivers of carbon price (Alberola et al. 2008).

Carbon price is apt to be greatly influenced by energy prices. As CO_2 emission is mainly resulted from fossil energy consumption, and power plants can selectively use various fuels such as coal, gas and oil, there is an internal price transmission mechanism between fossil energy market and carbon market. Therefore, carbon price is greatly influenced by energy prices: rising energy price is likely to cause the increase of carbon price, vice versa. This finding is consistent with that of Kanen (2006), Convery and Redmond (2007), Mansanet Bataller et al. (2007), Oberndorfer (2009), Hintermann (2010) and Mansanet-Bataller et al. (2011).

Carbon price is greatly affected by external heterogeneous environments. As a policy-based artificial market, carbon market is influenced by both the market mechanisms and external heterogeneous environments such as global climate negotiations, quotas allocation, financial crisis, and information pronouncements. In May 2006, the leakage of certified data induced a greatest fluctuation of carbon price; while the global financial crisis begun in September 2008, led to the sharp drop of carbon price: the price decreased from over 20 Euro/CO_2 equivalent to below 15 Euro/CO_2 equivalent. This is because the economic recession reduces the demands, which further results in the decreasing yields. In this case, the permits increase, which raises the supply of carbon market and the demands reduce. As a result, the carbon price decreases. This result is also verified by Christiansen et al. (2005), Zachmann and von Hirschhausen (2008), Chevallier et al. (2009), Mansanet-Bataller et al. (2011).

Carbon price is also sensitive to temperature conditions. Since 55% leaseholders of European emission allowances (EUA) are from thermal and electric departments, the shortage of EUA and rising carbon price appear owing to dry and cold winter calls for large amounts of heats which decrease the demands in hydropower; in the hot and dry summer, the demands for electricity largely grow, while there is shortage of hydropower resources. High temperature leads to the frequent main-tenance of nuclear power. Thus, power consumption based on coal makes CO_2 emission rise, and carbon price therefore grows. This outcome is also supported by Mansanet-Bataller et al. (2007), Alberola et al. (2008), Daskalakis et al. (2009), Benz et al. (2009), and Hintermann (2010).

Carbon price is remarkably subject to economic activities. Industrial production activities can directly determine the supply–demand relationship of EUA: the more economic activities, the more the more participants in carbon market, and the larger the demands of EUA, which giving rise to the increasing price, vice versa. This finding is also verified by Seifert et al. (2008), Chevallier (2009), Hintermann (2010).

It is noted that, the drivers of carbon prices of EU ETS at Phase I (2005–2007) and Phase II (2008–2012) are changed. Wei et al. (2010) used the co-integration technology to examine the interactions of carbon price and energy prices at the both long and short terms. They found that energy prices are slightly associated with carbon future price at Phase I, and has a long equilibrium relationship with carbon future price at Phase II. The variation of energy prices have been the main drivers of that of carbon price at Phase II. Keppler et al. (2010) adopted the Granger causality test to explore the relationship between carbon price and energy prices. They obtained that coal and gas prices at Phase I influenced carbon price, which further influenced electricity price; while at Phase II, gas price is still influenced, but coal price is no longer influenced by carbon price. Moreover, electricity price is a driver of carbon price at Phase II, in contrast at Phase I. Stock price becomes a driver to energy price at Phase II instead of being a followers of energy price at Phase I. In addition, temperature condition is an important driver of carbon price at both Phases I and II. Mansanet-Bataller et al. (2011) found out that energy price and information disclosure at Phases II are major drivers, while economic activities and temperature conditions are not primary divers of carbon price, same as the results at Phases I using the TGARCH model. Guebrandsdóttir and Haraldsson (2011) shown that CER price can preferably forecast EUA price, but EUA price is not significantly driven by electricity price. Creti et al. (2012) used the co-integration technology to compare the drivers at Phases I and II, and found that there existed two deferent long-term co-integration relationship between carbon prices and energy prices at Phases I and II, when considering the structural breaks in 2006.

1.2.2 Carbon Price Singlescale Forecasting

Pricing and forecasting carbon market is a hot focus, also is a challenge in the world. Although diverse approaches have been applied for dealing with this issue, they are roughly classified into five groups from the perspective of modeling.

(1) Market structure models: Starting from the market structure, general tendency of carbon price is explored by dynamic game analysis between market participants. Reilly and Paltsev (2005) built a EPPA-EURO model based EPPA developed by MIT for the tendency of carbon price of EU ETS, and obtained results showed that carbon price would be 0.6–0.9 Euro/CO_2 equivalent during 2005–2007, which is far smaller than real 20–25 Euro/CO_2 equivalent. One main reason may be attributed to the obvious defects in the model.

(2) Cost–benefit models: from the perspective of cost–benefit, minimum cost, and/or maximum benefit are obtained by seeking the optimal carbon price. Seifert et al. (2008) conducted a simulation analysis by constructing a stochastic computable equilibrium model of carbon price with the principle of minimum cost. However, the results obtained are not stable.

(3) Future market models: the price discovery function of carbon future market is used for carbon spot price forecasting. Keppler and Mansanet-Bataller (2010) performed a Granger causality test and found the efficiency of carbon future market at Phase I. while Montagnoli and deVries (2010) performed a variance ratio test and discovered that the carbon future market at Phase I are inefficiency. Feng et al. (2011) came to similar conclusions using a nonlinear approach. Chevallier (2010a), and Bredin and Muckley (2011) proposed that carbon future market in Phase II is inefficiency by using co-integration VAR model and likelihood ratio tests. This method is mainly used to test whether carbon future market is efficiency or not, and little used for forecasting carbon market.

(4) Multi-factor forecasting models: taking energy prices including coal, oil, gas and electricity prices, external heterogeneous environments indicated by virtual variables, temperature conditions, economic activity, etc., as the independent

variables, and carbon price as the dependent variable, a function relationship is built for forecasting carbon market. The methods used are divided into two types: linear and nonlinear approaches. Mansanet-Bataller et al. (2007) and Alberola (2008) employed multiple linear regression model (MLRM) containing virtual variables to explore carbon price, however they did not forecast carbon price. Guobrandsdottir and Haraldsson (2011) adopted a MLRM to conduct in-sample predication on actual carbon price at Phase II. Chevallier (2010b) carried out an in-sample predication on the fluctuation of carbon price at Phase II using the FAVAR model. Chevallier (2009) used different GARCH models to explore carbon price instead of predication.

(5) Time series forecasting models: carbon price past values contain all feature information, which can be used to deduce future values. Time series forecasting approaches can be divided into four groups: ① linear models. Chevallier and Sevi (2011) used the HAR-RV model to perform a recursive rolling forecasts on the carbon price at Phase II; ② nonlinear models. Paolella and Taschini (2008) used the GARCH model for forecasting the carbon price at Phase I. Benz and Truck (2009) used the MS-AR-GARCH model to perform a recursive rolling forecasts on the returns of carbon price at Phase I. Conrad et al. (2010) used the FIAPGARCH model to forecast the carbon prices at Phases I and II. ③ nonparametric models. Chevallier (2011c) applied the nonparametric modeling method to conduct a recursive rolling forecasts on the carbon price at Phases I and II, and the obtained results showed that it was superior to linear AR model. ④ artificial intelligence models. Zhu and Wei introduced LSSVM to forecast the carbon price, and their results showed that it could beat the ARIMA and ANN models. Moreover, Zhu (2012) used an EMD-based ANN model to predicate carbon price, and achieved a higher forecasting accuracy compared to other popular forecasting models.

1.2.3 Carbon Price Multiscale Forecasting

Fourier transform and wavelet analysis are the commonly used multiscale decomposition methods. However, the former is applicable to linear analysis, while the latter is incapable of adaptive decomposition subjected to the presetting of wavelet basis. Empirical mode decomposition (EMD) is an adaptive decomposition algorithm, which is more applicable to decompose nonstationary and nonlinear time series. EMD was initially applied in the fields of marine, natural science, and engineering (Huang et al. 1998). In recent years, EMD has been applied in the field of social science. Drakakis (2008) used the EMD to predict financial data. Zhang et al. (2008) and Yang et al. (2010) applied the EMD to forecast crude oil price. Yu et al. (2010) used the EMD to predict financial crisis. Zhu (2012) used the EMD-based ANN to forecast carbon price. Followed by Zhu (2012), in this book, EMD is also used for carbon price multiscale forecasting.

Once carbon price data are decomposed by EMD, a series of intrinsic mode functions (IMFs) and one residue with more stable, simpler structure, and stronger regular can be obtained, which are easily forecasted. IMF forecasting models can be classified into two groups as well: ① econometric models. Yu et al. (2008), Zhang and Wei (2010) applied ARIMA and GARCH to forecast IMFs, respectively. ② artificial intelligence-based nonlinear models. Yu et al. (2008) used ANN to forecast IMFs, while Wang et al. (2011), Zhu et al. (2016) used least squares support vector regression (LSSVR) to forecast the IMFs of hydropower consumption and nuclear power consumption in China, respectively. In this book, LSSVM is also used to forecast the IMFs of carbon price.

Multiscale ensemble forecasting is use to aggregate the forecasting values of the IMFs including the residue of carbon price into that of the original carbon price. Traditional multiscale ensemble forecasting approaches include three groups: ① direct aggregation approach: the sum of the forecasting values of the IMFs including the residue as that of the original carbon price, which is the mostly widely used multiscale ensemble forecasting approach; ② nonlinear aggregation approach: ANN, SVR and LSSVR (Yu et al. 2008; Tang et al. 2012) are used to integrate the forecasting values of the IMFs including the residue as that of the original carbon price. In this book, all the newly developed and these existing approaches are used for carbon price multiscale ensemble forecasting.

It is worth pointing out that, the popular evaluation criteria are used for carbon price level forecasting, including mean absolute error (MAE), mean absolute percentage error (MAPE) and root mean square error (RMSE) (Benz and Truck 2009; Chevallier 2010b; Chevallier 2011b). Few is used for carbon price directional forecasting such as (Zhu 2012) and Directional change accuracy (DCA) (Chen and Wang 2007). Few, except for Chevallier (2011a), performed the Diebold-Mariano test (Diebold and Mariano 1995), and/or White's Reality test (Wang and Yang 2010) to compare the abilities of various models for carbon price forecasting. In this book, all the newly developed and these existing evaluation criteria are used for carbon price ensemble forecasting.

1.3 The Organization of This Book

The organization of this book, as shown in Fig. 1.1, is in detail as follows:

In the chapter provides an accessible introduction to the importance, literature review and architecture of this book.

Chapter 2 explores the drivers of carbon price in the EU ETS during 2006 to 2012 using the structure breakpoint test, co-integration techniques and ridge regression. The empirical results show that the 2007s Bali action plan, 2008s global financial crisis and 2011s European debt crisis have significant effects on carbon price. Each effect has led to a structural breakpoint in the carbon price. Meanwhile, a cointegration relationship existed between carbon price and its drivers including energy prices, weather conditions, economic activities, and institutional decisions.

Fig. 1.1 The framework of
this book

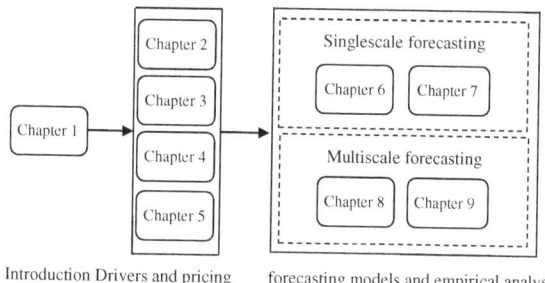

Equilibrium values show that the observed carbon price has been lower than its equilibrium value since October 2009. Carbon price will still tend to be depreciated in the future.

Chapter 3 examines the structural changes of European carbon futures price under the EU ETS during 2005–2012. More specifically, relying on a daily European allowance (EUA) futures contract, we investigate the structural changes of European carbon futures price using the structure breakpoint test: iterative cumulative sums of squares (ICSS) algorithm and event study models. The results show that since 2005, there have been three breakpoints of European carbon futures price, arisen from the two extreme events of 2008 global financial crisis and 2011 European debt crisis, which can contribute to understanding the pricing mechanism of carbon market and effectively forecasting carbon prices.

Chapter 4 examines the drivers of European carbon futures price under the EU ETS from a perspective of multiscale analysis. We extend the ensemble empirical mode decomposition (EEMD) to the EU ETS carbon price analysis. First, carbon price is decomposed into a number of independent intrinsic mode functions (IMFs) from high to low frequency and one residue via EEMD. Secondly, the fine-to-coarse reconstruction algorithm is employed to reconstruct the resulting decomposed IMFs into a high frequency component, a low frequency component and a trend component. The economic meanings of these three components are respectively identified as short-term fluctuations caused by normal supply–demand disequilibrium or some other market activities, the effect of a shock of a significant event, and a long-term trend. Finally, considering the characteristics of these three components, a principle of "divide and conquer" for carbon price forecasting is proposed.

Chapter 5 investigates the European carbon futures price dynamics by applying the Zipf analysis. Based on time series of European carbon futures, we analyze the information of price fluctuations with two parameters τ (speculator's timescale of investment) and ε (speculator's expectation of returns) by mapping the τ-returns of price into binary sequences (relative frequencies) and three-characteristic sequences (absolute frequencies), which contain the fundamental information of price fluctuations. Moreover, we explore the parameters empirically identified against various types of speculator's cognition patterns of price behaviors. Finally, we discuss the causes of formation of those cognition patterns and significant distortions of price behaviors. The results are helpful to hedge against unwanted price movements, and to understand the transactions between different types of agents.

Chapter 6 proposes a hybrid ARIMA and least squares support vector machine (LSSVM) model for carbon price forecasting. Due to the inherently high complexity, carbon prices simultaneously contain linear and nonlinear patterns. Although the traditional ARIMA model has been one of the most popular linear models in time series forecasting, the ARIMA model cannot capture the nonlinear patterns. LSSVM, a novel neural network technique, has been successfully applied in solving nonlinear regression estimation problems. Therefore, we propose a novel hybrid methodology that exploits the unique strength of the ARIMA and LSSVM models in forecasting carbon prices. Additionally, particle swarm optimization (PSO) is used to find the optimal parameters of LSSVM in order to improve the prediction accuracy. The empirical results obtained demonstrate that the proposed hybrid model can outperform the single ARIMA, LSSVM models, and the combined ARIMA and LSSVM (COMBINED) model in terms of statistical measures and trading performances.

Chapter 7 develops a parameters simultaneous optimization of phase space reconstruction (PSR) and LSSVM with uniform design for carbon price forecasting so as to obtain high forecasting accuracy and high modeling efficiency. First, as a large sample combinatorial optimization of multiple factors and multiple level in essence, the parameters simultaneous optimization of PSR and LSSVM is transformed into a small sample combinatorial optimization through uniform design, so as to enhance the optimization efficiency. Second, all the parameters are simultaneously optimized using the self-invoking LSSVM to obtain the optimal parameters. Finally, the proposed method is verified by forecasting two carbon futures prices with different maturities under the EUETS. The empirical results show that comparing with particle swarm optimization, the proposed method can significantly improve the modeling efficiency at the same time ensuring a high prediction accuracy, which seems an effectively approach for carbon price forecasting.

Chapter 8 proposes a multiscale prediction model hybridizing EMD, PSO and LSSVM to predict carbon price. First, carbon price is disassembled into a batch of high regular IMFs and a monotonic residue using EMD. Then, each component is independently predicted using LSSVR trained using PSO. Finally, the final forecasting results for all the IMFs and residue are aggregated into the final predictive values of original carbon price. To verify the proposed multiscale forecasting approach, two different-matured carbon futures prices under the EU ETS are used, and the obtained outcomes reveal that the presented approach can exceed the existing popular forecasting techniques.

Chapter 9 develops an adaptive multiscale ensemble learning paradigm incorporating EEMD, PSO, and LSSVM with kernel function prototype to forecast nonstationary and nonlinear carbon price. First, the extrema symmetry expansion EEMD, which can effectively restrain the mode mixing and end effects, is used to decompose carbon price into simple modes. Secondly, by using the fine-to-coarse reconstruction algorithm, the high frequency, low frequency and trend components are identified. Furthermore, ARIMA is applicable to predicting the high frequency components. LSSVM is suitable for forecasting the low frequency and trend components. At the same time, a universal kernel function prototype is introduced for making up the drawbacks of single kernel function, which can adaptively select the optimal kernel function type and model parameters according to the specific data using the PSO algorithm. Finally, the prediction results of all the components are aggregated into the forecasting values of original carbon price. The empirical results show that, compared with the popular prediction methods, the proposed method can significantly improve the prediction accuracy of carbon price, with high accuracies both in the level and directional predictions.

References

Alberola E, Chevallier J, Cheze B (2008) Price drivers and structural breaks in European carbon prices 2005–2007. Energy Policy 36(2):787–797

Benz E, Truck S (2009) Modeling the price dynamics of CO_2 emission allowances. Energy Econ 31(1):4–15

Bredin D, Muckley C (2011) An emerging equilibrium in the EU emissions trading scheme. Energy Econ 33:353–362

Chen KY, Wang CH (2007) A hybrid SARIMA and support vector machines in forecasting the production values of the machinery industrial in Taiwan. Experts Syst Appl 32:254–264

Chevallier J, Ielpo F, Mercier L (2009) Risk aversion and institutional information disclosure on the European carbon market: a case-study of the 2006 compliance event. Energy Policy 37:15–28

Chevallier J (2009) Carbon futures and macroeconomic risk factors: a view from the EU ETS. Energy Econ 31(4):614–25

Chevallier J (2010a) A note on cointegrating and vector autoregressive relationships between CO_2 allowance spot and futures prices. Econ Bull 30(2):1564–1584

Chevallier J (2011a) Wavelet packet transforms analysis applied to carbon prices. Econ Bull 31(2):1731–1747

Chevallier J (2010b) Volatility forecasting of carbon prices using factor models. Econ Bull 30(2):1642–1660

Chevallier J (2011b) Detecting instability in the volatility of carbon prices. Econ Econ 33:99–110

Chevallier J (2011c) Nonparametric modeling of carbon prices. Energy Econ 33(6):1267–1282

Chevallier J, Sevi B (2011) On the realized volatility of the ECX emissions 2008 futures contract: distribution, dynamics and forecasting. Ann Finance 7:1–29

Christiansen A, Arvanitakis A, Tangen K, Hasselknippe H (2005) Price determinants in the EU emissions trading scheme. Climate Pol 5:15–30

Conrad C, Rittler D, Rotfub W (2010) Modeling and explaining the dynamics of European Union Allowance prices at the high-frequency. Energy Econ 34(1):316–326

Convery FJ, Redmond L (2007) Market and price developments in the European Union emissions trading scheme. Rev Environ Econ Pol 1(1):88–111

Creti A, Jouvet PA, Mignon V (2012) Carbon price drivers: Phase I versus Phase II equilibrium? Energy Econ 34(1):327–334

Daskalakis G, Psychoyios D, Markellos, RN (2009) Modeling CO_2 emission allowance prices and derivatives: Evidence from the European trading scheme, J Bank Financ, 33(7):1230–1241

Diebold FX, Mariano RS (1995) Comparing predictive accuracy. J Bus Econ Stat 13(3):253–263

Drakakis K (2008) Empirical mode decomposition of financial data. Int Math Forum 3(25):1191–1202

Feng ZH, Zou LL, Wei YM (2011) Carbon price volatility: evidence from EU ETS. Appl Energy 88:590–598

Guobrandsdottir HN, Haraldsson HO (2011) Predicting the price of EU ETS carbon credits. Syst Eng Procedia 1:481–489

Hintermann B (2010) Allowance price drivers in the first phase of the EU ETS. J Environ Econ Manag 59:43–56

Huang NE, Shen Z, Long SR (1998) The empirical mode decomposition and the Hilbert spectrum for non-linear and non-stationary time series analysis. Proc R Soc Lond 454:903–995

Kanen JLM (2006) Carbon trading and pricing. Environmental Finance Publications, London

Keppler JH, Mansanet-Bataller M (2010) Causalities between CO_2, electricity, and other energy variables during phase I and phase II of the EU ETS. Energy Policy 38:3329–3341

Mansanet-Bataller M, Pardo AVE (2007) CO_2 prices, energy and weather. Energy J 28(3):73–92

Mansanet-Bataller M, Chevallier J, Herve-Mignucci M, Alberola E (2011) EUA and Scer phase II price drivers: Unveiling the reasons for the existence of the EUA–sCER spread. Energy Policy 9:1056–1069

Montagnoli A, deVries FP (2010) Carbon trading thickness and market efficiency. Energy Econ 32:1331–1336

Oberndorfer U (2009) EU emission allowances and the stock market: evidence from the electricity industry. Ecol Econ 68(4):1116–1126

Paolella MS, Taschini L (2008) An econometric analysis of emission allowance prices. J Banking Finan 32:2022–2032

Reilly JM, Paltsev S (2005) An analysis of the European emission trading scheme. Report No. 127, MIT Joint Program on the Science and Policy of Global Change

Seifert J, Uhrig-Homburg M, Wagner M (2008) Dynamic behavior of CO_2 spot prices. J Environ Econ Manag 56:180–194

Tang L, Yu L, Wang SY et al (2012) A novel hybrid ensemble learning paradigm for nuclear energy consumption forecasting. Appl Energy 93(5):432–443

Wang S, Yu L, Tang L, et al (2011) A novel seasonal decomposition based least squares support vector regression ensemble learning approach for hydropower consumption forecasting in China. Fuel Energy Abstr 36(11):6542–6554

Wang T, Yang J (2010) Nonlinearity and intraday efficiency tests on energy futures markets. Energy Econ 32:496–503

Wei YM, Wang K, Feng ZH et al (2010) Carbon finance and carbon market: models and empirical analysis. Science Press

Yangzhi (2010) How to realize the development of low carbon economy in China? http://discover.news.163.com/10/0201/10/5UE9C2P3000125LI.html

Yu L, Wang SY, Lai KK (2008) Forecasting crude oil price with an EMD-based neural network ensemble learning paradigm. Energy Econ 30:2623–2635

Yu L, Wang SY, Lai KK et al (2010) A multiscale neural network learning paradigm for financial crisis forecasting. Neurocomputing 73:716–725

Zachmann G, von Hirschhausen C (2008) First evidence of asymmetric cost pass-through of EU emissions allowances: examining wholesale electricity prices in Germany. Econ Lett 99 (3):465–469

Zhang X, Lai KK, Wang SY (2008) A new approach for crude oil price analysis based on empirical mode decomposition. Energy Econ 30:905–918

Zhang YJ, Wei YM (2010) An overview of current research on EU ETS: evidence from its operating mechanism and economic effect. Appl Energy 87(6):1804–1814

Zhu BZ (2012) A novel multiscale ensemble carbon price prediction model integrating empirical mode decomposition, genetic algorithm and artificial neural network. Energies 5:355–370

Zhu B, Shi X, Chevallier J et al (2016) An adaptive multiscale ensemble learning paradigm for nonstationary and nonlinear energy price time series forecasting. J Forecast 35(7):633–651

Chapter 2
European Carbon Futures Prices Drivers During 2006–2012

Abstract This chapter discusses the main driving factors behind carbon prices in detail. It presents key data, then proceeds with the results of cointegration test, Granger causality test, and ridge regression estimation. The chapter provides as well a comparative analysis of the equilibrium carbon price and observed carbon price, before drawing conclusions from the research.

2.1 Introduction

Global climate change is one of the most complex challenges facing people in the twenty-first century. To fulfill the commitments of the "Kyoto Protocol" at as low a cost as possible, the European Union Emissions Trading Scheme (EU ETS) was established by EU Directive 2003/87/EC in 2003 and started in January 2005. The EU ETS set CO_2 emissions upper limits for 12,000 emission facilities, such as generator sets, oil refining equipments, building materials, paper-making equipments, and metal manufacturing equipments, across 25 EU member states. The EU ETS can be divided into four phases: Phase I (2005–2007), Phase II (2008–2012), Phase III (2013–2020), and Phase IV (2021–2028). After years of rapid development, the EU ETS has gradually developed into a financial market covering carbon spot, futures, options, and other trading products. Whether in market value or trading volume, the EU ETS is currently the world's largest carbon market. Its value is much higher than other major global carbon markets, and also significantly exceeds the clean development mechanism (CDM) carbon market. Moreover, the EU ETS is a weather vane for global carbon market trading. Its development shapes the direction of global carbon market, and its market situation directly affects the reference prices for global CO_2 trading (Zhang and Wei 2010). In recent years, the EU ETS has become an important tool as mankind tries to cope with climate change, as well as a major choice for investors diversifying their investment risks. Therefore, the issues surrounding carbon market and carbon finance have become foci for energy and climate change researchers (Wei et al. 2010).

© Springer International Publishing AG 2017
B. Zhu and J. Chevallier, *Pricing and Forecasting Carbon Markets*,
DOI 10.1007/978-3-319-57618-3_2

In recent years, more and more researchers around the world began to pay attention to the EU ETS carbon market. Considering the environmental benefit and cost-efficiency of this EU ETS carbon market, many researchers have studied its Phase I to date, although this phase is, in essence, a learning phase. For instance, Mansanet-Bataller and Pardo (2007) and Alberola et al. (2008, 2009) inspected the driving factors of 2005–2007 carbon prices successively. Paolella and Taschini (2007), Daskalakis (2008), Seifert et al. (2008), Benz and Truck (2009) explored the carbon price behaviors in Phase I for prediction. However, only a few researchers, such as Feng et al. (2011), Chevallier (2012), and Creti et al. (2012) studied the drivers of carbon prices in Phase II.

Existing research results can provide this study with important references. However, they also show some disadvantages. First, existing studies mainly focus on the EU ETS Phase I. Since the market experiences, market characteristics (mobility and depth), and market rules of Phase II differ from those of Phase I, the results of Phase I may not apply to Phase II, particularly the driving factors used in testing carbon prices. Second, existing results are basically obtained from carbon spot prices. At present, carbon spot trading is still very low, while carbon futures trading are the main product in carbon markets. Besides, carbon futures contracts can also yield higher research value (World Bank 2012). However, carbon futures price studies are rare. Since carbon futures enjoy greater trading volume than carbon spot price trades theoretically, carbon futures prices are much less sensitive to the important structural changes that have occurred on the spot market during the study period—January 2006 to April 2012. Thus, carbon futures prices show more steady fluctuation than spot price equivalents. Thereby, the research conclusions from carbon spot prices are probably not applicable to carbon futures prices. This situation is not beneficial when trying to grasp a general view of the driving factors of carbon prices and cannot provide investment decision-makers with enough information supports. Third, existing studies mainly employ traditional multiple linear regression method. This method can basically cause multicollinearity in existing results. Therefore, the reliability of research results is low, and the driving factors of carbon prices cannot be effectively grasped. In summary, although the EU ETS has attracted the attention of researchers for several years, its study is still in the initial stage.

This study is designed to explore the driving factors of carbon futures price over the Phase I and Phase II–January 2006 to April 2012. Being similar to the study by Creti et al. (2012), we use the cointegration techniques to identify the determinants of the carbon price over the whole study period. This study extends the study of Creti et al. (2012) in two aspects: first, more driving factors are considered in this study, i.e., energy prices are imported with crude oil price, as well as coal, gas, and electricity prices; besides temperature conditions, economic activities and institutional decisions which intensively influence the EU ETS carbon market, are introduced into determine the driving forces of carbon futures price. As far as we know, the influences of 2007s Bali action plan, 2008s global financial crisis, and 2011s European debt crisis on the EU ETS carbon market have not been empirically analyzed. Thus, in this chapter, we seek to measure the time-points of structural changes of carbon price series probably caused by these events using the BP structure breakpoint test algorithm proposed by Bai and Perron (2003). Furthermore, these time-points are included amongst the driving factors. Second, to eliminate multicollinearity among the independent variables to obtain more reliable regressive results, the ridge regression method is used to deduce the equilibrium carbon price and reveal the main reasons for the difference between the equilibrium carbon price and the observed carbon price.

Our results show that 2007s Bali action plan, 2008s global financial crisis, and 2011s European debt crisis all exerted significant influences on carbon prices. Each influence leads to a structural breakpoint of carbon price; meanwhile, a long-term cointegration relationship existed between carbon price and its driving factors including energy prices, weather conditions, economic activities and institutional decisions; equilibrium values show that the observed carbon price has been lower than its equilibrium value since October 2009, and carbon price still tends to be depreciated in the future.

2.2 Carbon Price Drivers

Key factors such as energy prices, weather conditions, economic activities, and institutional decisions can exert significant impacts on carbon price (Alberola et al. 2008).

Energy prices exert obvious influences on carbon price. Since fossil energy consumption is the main source of CO_2 emissions, power enterprises can switch within various fossil fuels-coal, natural gas, or oil. Thus an internal price transmission mechanism is induced between fossil energy and carbon markets. Carbon price is thus closely connected to energy prices. The rising in energy prices will induce the rise of carbon price, while energy prices fall will also cause decreased carbon price. This idea is supported by Kanen (2006), Convery and Redmond (2007), Mansanet-Bataller and Pardo (2007), Oberndorfer (2009), Hintermann (2010), Mansanet-Bataller et al. (2011).

Carbon market is a temperature-sensitive market, so temperature conditions significantly influence carbon price. Since approximately 55% of the EU allowance (EUA) holders are operating in the heat or electricity sectors, in cold and dry winters, more demand for heat, and less output for hydropower can cause the shortage of EUA and rising carbon price; in hot and dry summers, the surge in electricity demand causes a shortage of hydropower resources. Meanwhile, nuclear power is frequently maintained due to high temperatures. Therefore, electricity supplies rely on coal, resulting in the growth of CO_2 emissions and their cost. This idea is supported and checked by Mansanet-Bataller and Pardo (2007), Alberola et al. (2008), Daskalakis 2008), Benz and Truck (2009), Hintermann (2010), Wei et al. (2010).

Economic activities show obvious influences on carbon price. Industrial production activities directly determine EUA's supply and demand. An increase in economic activities can draw in more market participants and produce more demands, thus carbon price will rise; conversely, a reduction in economic activities will cause a reduction in the number of market participants: demand and carbon price thus fall. Seifert et al. (2008), Hintermann (2010), Chevallier (2012) support and validate this idea.

Institutional decisions exert significant influences over carbon price. As a policy product of CO_2 emissions reduction protocols for EU member states, carbon price is affected by market mechanisms as well as external heterogeneous environments. Some institutional decisions, such as international climate negotiations, allowance allocation, financial crisis, and important announcements, can influence carbon price and cause large fluctuations therein. For example, due to the influence of certified data leakage event in May 2006, carbon price showed a much larger decrease. The global economic crisis that started in September 2008 caused carbon price to drop from 20 €/t to 15 €/t. Economic recession greatly dampens demand, thus output reduces and demand for EUA substantially decrease, resulting in the increase of carbon market supply, demand reduction, and subsequently lower carbon price. This conclusion is well supported by Christiansen et al. (2005), Zachmann and von Hirschhausen (2008), Alberola et al. (2009), Chevallier et al. (2009), Mansanet-Bataller et al. (2011).

Owing to its variations over different EU ETS phases, carbon price presents complex relationships with energy prices, weather conditions, economic activities, institutional decisions, and other factors. Wei et al. (2010) examined the long-term and short-term interactions between the EU ETS carbon price and energy prices using cointegration techniques. The results showed that energy prices displayed a weak relationship with carbon futures price in Phase I, but presented a long-term equilibrium relationship with carbon futures price in Phase II. Energy prices variations were also an important cause behind carbon price changing in Phase II. Keppler and Mansanet-Bataller (2010) studied the relationship of carbon price and energy prices using Granger causality test method. They found that in Phase I carbon price was affected by coal and natural gas prices, then carbon price influenced electricity price; in Phase II, natural gas was still an important factor influencing carbon price, but coal price no longer influenced carbon price, and carbon price was also no longer a factor influencing electricity price. On the contrary, electricity price influenced carbon price; stock price was transformed from an energy price-follower in Phase I to a price-driver in Phase II; weather conditions showed important influences on carbon price in both Phases I and II. Mansanet-Bataller et al. (2011) found that energy prices were the main driving force in Phase II using their TGARCH model, while economic activities and temperature conditions were no longer significant factors. This conclusion was different from that of Phase I. Guebrandsdóttir and Haraldsson (2011) found that certified emissions reductions (CERs) price could well predict EUA price, and electricity price did not significantly affect EUA price when studying carbon price prediction. Creti et al. (2012) compared carbon price driving factors of Phases I and II by cointegration techniques. They concluded that there were different long-term cointegration relationships between carbon price and energy prices of the two phases considering 2006s structural breakpoint.

2.3   Data

2.3.1   Carbon Price

The European Climate Exchange (ECX) is the largest carbon exchange in the EU ETS system. The daily carbon trading volume of this exchange accounts for more than 80% of the total carbon trading amount of the EU's main carbon exchanges. Therefore, its trading position largely reflects general trends in EU ETS carbon trading. This chapter selected monthly price data for EUA carbon futures contract matured in December 2012, namely DEC12, over the period January 2006 to April 2012. With €/t CO_2 as its unit, a total of 76 monthly data points was used: the carbon price was denoted by *Carbon*.

2.3.2   Energy Prices

The selected energy prices from January 2006 to April 2012 are indicated as follows: (1) oil price; ICE (intercontinental exchange) monthly Brent oil price was used in units of U.S. dollars per barrel, and denoted by Brent; (2) coal price; The monthly coal futures contract price at three influential harbors (Amsterdam, Rotterdam, and Antwerp) on the European Energy Exchange (EEX), located in Germany, was applied in units of €/t, and was denoted by Coal; (3) natural gas price. Since UK is the biggest natural gas consumer in the EU, ICE natural gas price movements basically shape the condition of the European natural gas market. Thus, the ICE monthly British Gas futures index price was used in this chapter in pence/British thermal unit (BTU), and was denoted by Gas; (4) electricity price; Germany's installed capacity and power generation rank first in Europe and Germany has the EU's largest electricity market. Thus, we selected the EEX electricity futures price, in units of €/MWH which was denoted by *Elec*.

2.3.3   Temperature Conditions

Tendances Carbone's EU temperature index monthly mean was used and denoted by *Temp*.

2.3.4 Economic Activities

Tendances Carbone's Europe industrial production index (seasonally adjusted) was introduced to reflect general European economic activities, and was denoted by *Indu*. The data selected in this chapter are presented in Table 2.1.

2.3.5 Institutional Decisions

To explore the structural changes in carbon price series, the time-points of structural changes therein were measured using global minimisation of the absolute sum of squared residuals (SSR) in the BP structure breakpoint test algorithm proposed by Bai and Perron (2003). Assuming that there are m breakpoints, the corresponding model is given by formula (2.1)

$$
\begin{aligned}
y_t &= x_t' \beta + z_t' \delta_1 + u_t, t = 1, 2, \ldots, T_1 \\
y_t &= x_t' \beta + z_t' \delta_2 + u_t, t = T_1 + 1, T_1 + 2, \ldots, T_2 \\
y_t &= x_t' \beta + z_t' \delta_{m+1} + u_t, t = T_m + 1, T_m + 2, \ldots, T.
\end{aligned}
\tag{2.1}
$$

where y_t is the value of the dependent variable at t, $x_t(p \times 1)$ and $z_t(p \times 1)$ are covariance vectors, β, δ_j, $j = 1, 2, \ldots, m+1$ are corresponding coefficient vectors; u_t is a random disturbance term, and $(T_1, T_2, \ldots, T_m)$ are breakpoints. Then calculate the minimum SSR values after one, two, and m structural change times within the sampling period respectively. Finally, Bayesian information criteria (BIC) were used to find the number of optimal breakpoints and each breakpoint's occurrence time.

Three breakpoints were found by this breakpoint test: September 2007, October 2008, and May 2011. These three breakpoints caused one surge, and two slumps, in carbon price, as shown in Figs. 2.1 and 2.2. Every surge and slump in carbon price is closely connected with institutional decisions. The first surge in the carbon price

Table 2.1 Descriptive statistics

	Mean	Range	Maximum	Minimum	Standard deviation	Skewness	Kurtosis
Carbon	18.46	23.73	30.82	7.09	5.82	0.22	−0.42
Brent	83.69	92.23	135.73	43.50	23.35	0.41	−0.81
Gas	5.72	6.97	9.45	2.48	2.01	−0.04	−1.35
Coal	76.72	90.59	135.56	44.97	27.99	0.64	−1.04
Elec	53.25	81.09	109.40	28.31	15.60	0.98	1.54
Temp	11.20	22.50	22.90	0.40	6.03	−0.02	−1.20
Indu	103.51	24.20	114.00	89.80	7.36	−0.40	−0.94

Fig. 2.1 Number of carbon price BP breakpoints (January 2006 to April 2012)

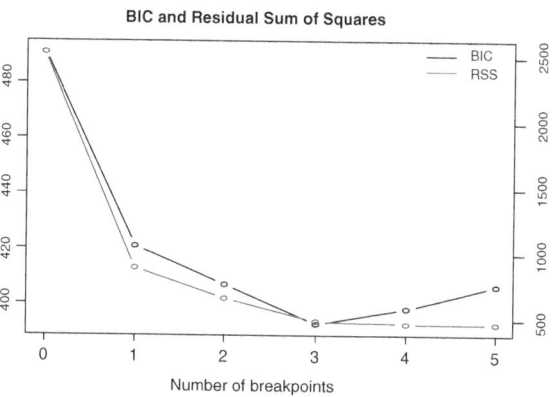

Fig. 2.2 Time-points of carbon price BP breakpoints (January 2006 to April 2012)

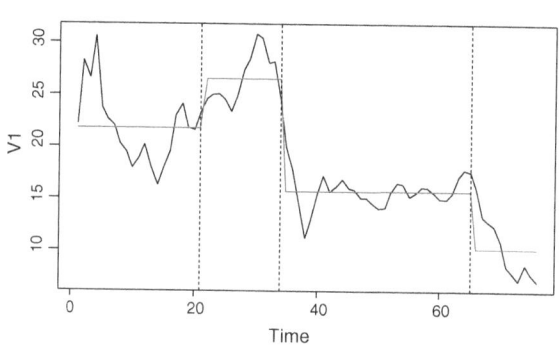

was derived from the implementation of more stringent emissions reduction policy announced by both globally and EU-wide. The first slump in the carbon price originated from the global financial crisis. The second slump was caused by the European debt crisis. Therefore, three dummy variables were introduced in this chapter to shape the impacts of these institutional decisions on the carbon market. The institutional decisions before, including, and after September 2007 were taken as 0 and 1 respectively and were denoted by Break1; those before, including, and after October 2008 were taken as 0 and 1 respectively, and were denoted by Break2; those before, including, and after May 2011 were taken as 0 and 1 respectively, and were denoted by Break3.

2.4   Cointegration Test and Ridge Regression Results

2.4.1   Cointegration Test

In this chapter, cointegration techniques were used to identify the potential long-term equilibrium relationship between carbon price and its driving factors.

Table 2.2 Unit root testing results

	ADF value	p-value		ADF value	p-value
Carbon	−1.518578	0.5188	ΔCarbon	−8.024510	0.0000***
Brent	−2.760113	0.0691*	ΔBrent	−4.908595	0.0001***
Gas	−2.153741	0.2248	ΔGas	−8.228513	0.0000***
Coal	−1.883104	0.3384	ΔCoal	−5.645959	0.0000***
Elec	−3.743625	0.0052***			
Temp	−9.484851	0.0000***			
Indu	−1.733334	0.4104	ΔIndu	−3.655754	0.0068***

Δ indicates first-order difference. ***(resp. **, *) indicates the rejection of the null hypothesis of a unit root at the 1% (resp. 5, 10%) significance level

First, the Augmented Dickey–Fuller (ADF) unit root test was applied to test the stability of those variables studied.

It can be observed from Table 2.2 that, except for several series such as oil price, electricity price and temperature condition were steady at the 10% significance level, the remaining series were first-order steady. According to econometric theory, it can be considered that all series were first-order steady. Therefore, a multiple linear regression model concerning carbon price and energy prices, temperature condition, economic activity and institutional decisions, can be obtained as formula (2.2)

$$Carbon_t = \beta_0 + \beta_1 Brent_t + \beta_2 Coal_t + \beta_3 Gas_t + \beta_4 Elec_t + \beta_5 Temp_t + \beta_6 Indu_t$$
$$+ \beta_7 Break1_t + \beta_8 Break2_t + \beta_9 Break3_t + \varepsilon_t,$$

$$(2.2)$$

where t refers to month t in the study period, and ε is an error term.

Second, Johansen's cointegration techniques were used to determine whether or not the multiple linear regression models above could be regarded as a long-term equilibrium relationship. Test results are shown in Table 2.3 which suggested that carbon price and its driving factors reject the null hypothesis containing no

Table 2.3 Johansen's cointegration trace test results (p-value)

Null hypothesis	Trace Statistic	Prob.**
None*	393.5157	0.0000*
At most 1*	283.1781	0.0000*
At most 2*	213.7940	0.0000*
At most 3*	149.1962	0.0008*
At most 4*	100.0302	0.0246*
At most 5	64.02284	0.1329

Trace test indicates 5 cointegration eqn(s) at the 0.05 level
*Denotes rejection of the hypothesis at the 0.05 level
**MacKinnon-Haug-Michelis (1999) p-values

cointegration relationship at the 5% significance level. This finding was in accordance with that by Creti et al. (2012) and the results from their study of Phase II by Bredin and Muckley (2011). However, it differed from that of Phase I by Bredin and Muckley (2011), in that, when the structural change in the carbon price series caused by the certified data leakage event in May 2006 was not considered, cointegration relationship did not exist in carbon price and its driving factors. The reason is possibly that Phase I of the EU ETS, as a highly uncertain emerging future market, may just have been a short trial period; when the structural change of carbon price series caused by the aforementioned May 2006 events were considered, a cointegration relationship reappeared between carbon price and its driving factors. This finding proved the conclusions matched those of Gregory et al. (1996): structural changes in time series were the preference of cointegration test, resulting in the error that: "the null hypothesis that should be rejected is denied due to the structural changes".

2.4.2   Ridge Regression Estimation

Based on Johansen's cointegration test results, the cointegration relationship between carbon price and its driving factors was estimated. First, the correlation between variables was evaluated by multicollinearity test: the results are shown in Table 2.4 where a strong significant correlation may be seen between some variables. Thus, it can be deduced that the correlation between some of the variables are strong and that there may be significant multicollinearity.

Second, ordinary least squares (OLS) estimation was introduced to further check whether or not there was multicollinearity. The OLS estimation results are shown in Table 2.5 which shows that OLS coefficients displayed extreme value phenomenon. Some coefficients were highly significant, while others were insignificant. Moreover, R square and the F-statistic indicated higher regression significance. Thus, it can be judged that multicollinearity was possibly present. Some inflation factors (VIFs) of independent variables were greater than 5, even 10, indicating that there was severe multicollinearity between variables. Due to the presence of this severe multicollinearity, the OLS estimation coefficients cannot be guaranteed to have been reliable. Therefore, the OLS estimation results cannot be used to conduct analysis and exercise judgement thereon. Only by eliminating this multicollinearity between the independent variables, can robust regression results be obtained.

Third, to overcome this multicollinearity between independent variables to obtain robust regression results, ridge regression was used to conduct model estimation. Ridge regression is a biased estimation regression method for data analysis in the face of multicollinearity: it is actually modified from OLS. It is a regression equation obtained by abandoning the non-biased nature of OLS, while sacrificing

Table 2.4 Correlation test results

	Carbon	Brent	Gas	Coal	Elec	Temp	Indu	Break1	Break2	Break3
Carbon	–									
Brent	−0.030	–								
Gas	0.041	0.673**	–							
Coal	−0.067	0.878**	0.806**	–						
Elec	0.438**	0.242*	0.593**	0.386**	–					
Temp	0.239*	0.098	−0.139	0.026	−0.178	–				
Indu	0.625**	0.279*	0.391**	0.284*	0.339**	0.090	–			
Break1	−0.336**	0.453**	0.373**	0.521**	0.214	−0.129	−0.424**	–		
Break2	−0.774**	0.103	0.038	0.199	−0.177	−0.155	−0.805**	0.682**	–	
Break3	−0.570**	0.567**	0.451**	0.577**	−0.068	−0.012	−0.053	0.259*	0.379**	–

** (resp. *) is significant at the 5% (resp. 10%) level (2-tailed)

Table 2.5 Ordinary least squares estimation results

	Unstandardized coefficients	t-Statistic	Sig.	VIF
C	16.821	1.512	0.135	
Brent	0.022	0.777	0.440	5.650
Gas	−0.281	−0.947	0.347	4.622
Coal	0.035	1.100	0.275	10.173
Elec	0.111	4.222	0.000*	2.184
Temp	0.165	3.313	0.002*	1.184
Indu	−0.041	−0.387	0.700	7.983
Break1	0.952	0.777	0.440	3.841
Break2	−7.769	−4.690	0.000*	8.912
Break3	−6.658	−5.635	0.000*	2.454
R Square	0.922			
F-Statistic	41.832			
Sig.	0.000			

*Is significant at the 1% level (2-tailed)

some information and accuracy, yet providing a regression coefficient more suitable for practical application. The residual standard deviation of a ridge regression is larger than that of an OLS regression, but it presents much stronger stability and pathological tolerance compared to OLS. The Gauss–Markov theorem indicates that multicollinearity does not affect estimators and minimum variance of an OLS. Though OLS estimation shows the minimum variance in all linear unbiased estimations, this minimum variance is not necessarily small. In fact, a biased estimation can be applied. Though a biased estimation may show a slight deviation, its accuracy is much higher than the unbiased estimation. Based on this principle, ridge regression estimation is obtained by introducing biased constants in the normal equations. Ridge regression is defined as $\beta(k) = (X'X + kI)^{-1}X'Y$. Where, X is the independent variable matrix, X' is the transpose of X, Y is dependent variable vector, k is the ridge parameter or biased parameter, usually such that $0 < k < 1$, and $\beta(k)$ is the ridge regression estimator of regression coefficients vector. When $X(t)$, ridge regression is reduced to OLS regression (Hoerl and Kennard 1970).

By observing the ridge tracks and changing tendency of ridge regression's estimated R square with k, the acquired regression coefficients of independent variables were more stabilized when $k = 0.10$. Therefore, $k = 0.10$ was selected for this ridge regression estimation: key results are shown in Table 2.6 where it can be seen that the ridge regression coefficients of most variables are significant at 10%, or even 5 and 1% levels. Meanwhile, R square reached 0.917, which indicated that the overall fit was good. Besides, the F-statistic passed the 1% significance level test, and the VIF of each independent variable was obviously less than 5. Therefore,

Table 2.6 Ridge regression estimation results ($k = 0.10$)

	Unstandardized coefficients	t-Statistic	Sig.	VIF
C	2.730857	0.560471	0.288528	
Brent	0.024816	1.578961	0.059563[*]	1.650
Gas	−0.185543	−0.986962	0.163633	1.747
Coal	0.008783	0.645156	0.260531	1.778
Elec	0.105719	5.261444	0.000001[***]	1.204
Temp	0.166533	3.719303	0.000207[***]	0.893
Indu	0.097748	2.023594	0.023532[**]	1.546
Break1	0.478986	0.599370	0.275489	1.537
Break2	−5.187872	−7.255254	0.000000[***]	1.559
Break3	−6.278589	−7.143728	0.000000[***]	1.275
R Square	0.917			
F-Statistic	38.824			
Sig.	0.000			

[***](resp. [***],[**]) is significant at the 1% (resp. 5, 10%) level (2-tailed)

each variable regression coefficient was basically in line with economic tests, and the overall model fitting effect was consistent with the prevailing EU ETS carbon market trading situation and its price driving mechanism.

2.4.3 Granger Causality Test

Based on the estimation results obtained above, Granger causality test results are shown in Table 2.7. The results showed that, over the whole period, carbon price was affected by energy (oil, gas, coal, and electricity) prices, as well as economic activities and institutional decisions (Break2 and Break3) in either the short, or long terms. The short-term or long-term impacts of institutional decision (Break1) and temperature condition on carbon price were not obvious. The results verified the relevant conclusions by Alberola et al. (2008), Keppler and Mansanet-Bataller (2010), Mansanet-Bataller et al. (2011), Creti et al. (2012). Meanwhile, it provided a stronger economic basis for these studies. Another interesting feature was that carbon price can generate short-term and long-term impacts on oil, gas, and electricity prices. This meant that the barriers between the EU ETS carbon market and energy markets such as those for oil, natural gas, and electricity, have been gradually eliminated by the information transmission mechanism. Therefore, various market prices present an incipient interaction, which enhances the status and role of carbon market in the macro-economy system.

Table 2.7 Granger causality test results (p-value)

	Carbon	Brent	Gas	Coal	Elec	Temp	Indu	Break1	Break2	Break3
Carbon	–	0.0651*	0.0829*	0.2231	0.0578*	0.2780	0.4361	0.9924	0.3954	0.8259
Brent	0.0142**	–	0.0002	0.0005	0.0232	0.4023	0.0055	0.7675	0.0173	0.0689
Gas	0.0184**	0.7633	–	0.8914	0.2987	0.1330	0.1891	0.9821	0.0700	0.2685
Coal	0.0371**	0.9523	0.0015	–	0.0775	0.3950	0.0208	0.7820	0.0800	0.0342
Elec	0.0095***	0.1445	0.2313	0.2605	–	0.1422	0.0798	0.4712	0.0074	0.9912
Temp	0.2920	0.1137	0.7453	0.4171	0.0309	–	0.6151	0.5437	0.3018	0.2427
Indu	0.0701*	0.9391	0.1148	0.7933	0.1163	0.9553	–	0.7096	0.4044	0.9996
Break1	0.5361	0.6092	0.2409	0.3691	0.0608	0.5496	0.1921	–	0.1486	0.7712
Break2	0.0182**	0.2902	0.2913	0.2183	0.0043	0.5287	0.0482	1.0000	–	0.5392
Break3	0.0677*	0.2465	0.4957	0.8579	0.6897	0.5236	0.9410	1.0000	1.0000	–

Note The test is based on the null hypothesis that the variable X in line does not cause the variable Y in column. *** (resp. **, *) indicates the rejection of the null hypothesis of no causality at the 1% (resp. 5, 10%) significance level. The p-values in parentheses are relating to short-run causality, the other reported p-values correspond to long-run causality

2.5 Equilibrium Carbon Price

2.5.1 Equilibrium Carbon Price Equation

According to the cointegration relationship between carbon price and its driving factors obtained above, the corresponding equilibrium carbon price can be deduced. Relying on the ridge regression estimation results, the equilibrium carbon price can be expressed as formula (2.3)

$$\text{Carbon}_t = 2.7309 + 0.0248\,\text{Brent}_t - 0.1855\beta_2\,\text{Coal}_t + 0.0088\,\text{Gas}_t + 0.1057\,\text{Elec}_t + 0.1665\,\text{Temp}_t$$
$$+ 0.0977\,\text{Indu}_t + 0.4790\,\text{Break1}_t - 5.1879\,\text{Break2}_t - 6.2786\,\text{Break3}_t$$

$$(2.3)$$

Natural gas and coal prices coefficients were not significant at the 10% level. This result contradicted Alberola et al. (2008), Keppler and Mansanet-Bataller (2010) and Chevallier (2012) who believed that changes in carbon price were derived from those of natural gas and coal prices. The reason for this contradiction may have been that: (1) Alberola et al. (2008) used OLS to analyze the driving factors of carbon spot price in Phase I, thus there may be multicollinearity. (2) This may explain that there are some barriers in the internal transmission mechanism, in particular, that of information, between carbon market and energy markets for natural gas, coal, etc. The information transmission between these two markets is not smooth enough. We have reasons to believe that with the gradual improvement of carbon market, natural gas, and coal prices would significantly affect carbon price.

First, of those significant energy prices, oil and electricity prices showed positive effects on carbon price. Electricity price coefficient was positive and significant at the 1% level. The power generation facilities of EU states are restricted due to a general lack of water resources. In EU member states, especially Germany, coal is used extensively for power generation to meet the demand for electricity. Large amounts of coal consumption will lead to increased CO_2 emissions, followed by an increased CO_2 allowance. Therefore, large power plants invoke the most important CO_2 emissions allowance demands. Electricity price rises with increased power generation cost mainly caused by the rising costs of coal and natural gas, carbon price is thereby increased. Oil price manifests a positive effect on carbon price. Specifically, when oil price rises, people tend to use relatively cheaper coal, thus more CO_2 is emitted and more CO_2 emissions allowance is needed, promoting the increase of carbon price; when oil price falls, people will generally reduce their use of coal since oil is cleaner than coal due to the lower CO_2 emissions coefficient of oil, meanwhile CO_2 emitted is reduced, and carbon price is thereby lowered.

Second, temperature condition showed positive effects on carbon price. Temperature condition coefficient was positive and significant at the 1% level. This ran contrary to the conclusions of Alberola et al. (2008) and Mansanet-Bataller et al. (2011). They considered that the temperature condition did not significantly influence carbon price, the reason may be that: (1) Temperature condition shows a nonlinear, rather than a linear, effect on carbon price. (2) Seasonal factors show more effects on carbon price than temperature condition. In theory, climate deterioration or extreme weather events' occurrence can exert entity impacts on carbon-regulated industries. These impacts are indirect and complex. In cold and dry winters, or in hot and dry summers, electricity demand surges. The increase in coal consumption can cause increasing CO_2 emissions, which promotes an increased carbon price. Moreover, drought affects hydropower generation to some extent and coal consumption will thence increase, leading to increasing CO_2 emissions and subsequent carbon price rises.

Third, economic activity showed positive effects on carbon price. Economic activity coefficient was positive and significant at the 1% level. This result was consistent with the conclusion of Creti et al. (2012), namely, economic activity showed significantly positive effects on carbon price, while it differed from the conclusion by Mansanet-Bataller et al. (2011). They argued that economic activity could not show positive effects on carbon price. The improvement of macro-economic situation will induce expansion of industrial production, which creates new demands for energy consumption such as electricity, coal, etc., and especially increases demand in high energy-consuming industries such as non-ferrous metallurgy, chemical, electrical, etc. These demands can enlarge the uptake of various energy sources, leading to increased CO_2 emissions. In the event of certain national emissions reduction allowances, new demand for CO_2 emissions allowances will be created to offset increasing CO_2 emissions arising from industrial production increases. Finally, carbon price increases are promoted. Otherwise, when macro-economic effects present a downtrend, energy consumptions will also come down significantly, resulting in the reduction of CO_2 emissions and an eventual carbon price drops.

Fourth, of those three structural breakpoints, the first (Break1) was not significant at the 10% level, while the following two (Break2 and Break3) showed negative impacts on carbon price, and were both significant at the 1% level. This result was inconsistent with the conclusions of Chevallier et al. (2009) who found that the certified information leakage event occurred in May 2006 significantly affected carbon price. The main reason may be that they used carbon spot price in Phase I, while we used the mixed carbon futures price of Phases I and II. BP structure breakpoint test results showed that this event did not cause structural changes in carbon price. The global financial crisis in 2008 and Europe's debt crisis in 2011 exerted stronger influences on carbon price than Bali Action Plan in 2007. The former two caused declining carbon price with an average amplitude of −5.1879 €/t and −6.2786 €/t respectively, which were both higher than the rising average amplitude of 0.4790 €/t caused by the latter. This may be the main reason that the former two were significant throughout the period, while the latter was insignificant.

2.5.2 Comparison of Observed Carbon Price and Equilibrium Carbon Price

To analyze whether or not the observed values over the whole period lay close to corresponding equilibrium values, we calculated the difference between the observed carbon price and equilibrium carbon price to reveal the carbon market price formation mechanism and stress the importance of carbon price prediction.

The observed and equilibrium carbon prices are shown in Fig. 2.3. The corresponding deviations are shown in Fig. 2.4. In general, the deviations were smaller, and the relative error concentrated to within 10%. However, those deviations in April 2006, October 2006, February 2007, May 2011, and November 2011 were much larger, with relative errors greater than 20%: in particular, in April 2012, it even approached 40%.

The appreciation defined in this chapter refers to the fact that the observed value is higher than its equilibrium value. Otherwise, it will represent depreciation. Thus it can be seen that appreciation and depreciation periods appeared alternately with carbon price changes from January 2006 to April 2012.

The first appreciation period occurred in the first half of 2006. During this period, the increase in energy prices and CO_2 emissions predictions, as well as traders' high expectations for carbon price, trigged carbon price inflation. If carbon price was determined by the balance of market supply and demand, it should not have reached such a high level. In fact, rising energy prices encouraged companies to substitute natural gas for coal to reduce CO_2 emissions. Therefore, carbon price fell.

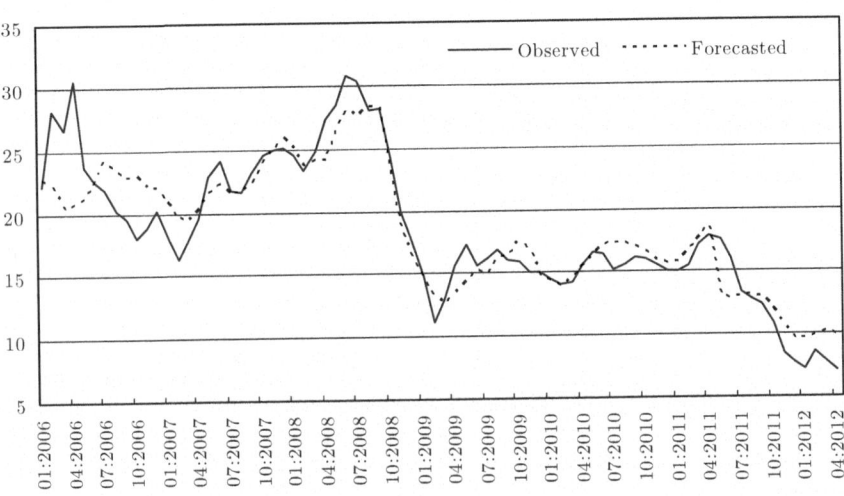

Fig. 2.3 Observed and equilibrium carbon prices

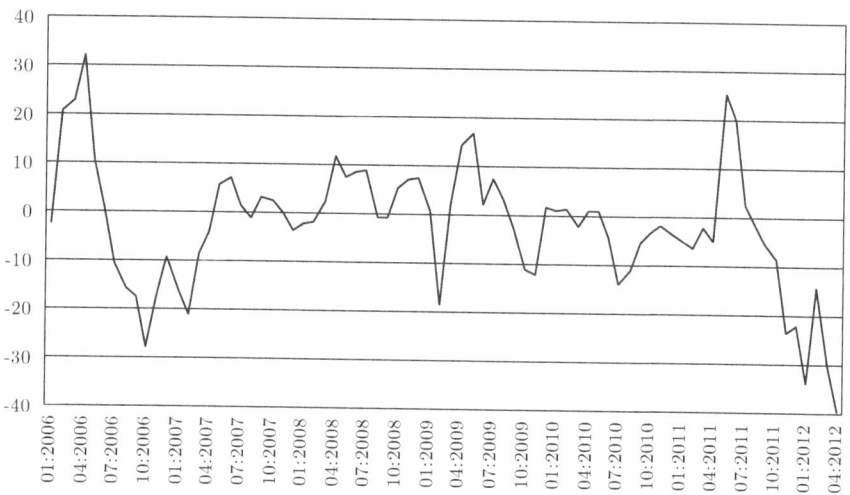

Fig. 2.4 Relative error between observed and equilibrium carbon prices

The first depreciation period occurred in the second half of 2006. Affected by the certified information leakage event occurred in May 2006, and EUA over-allocation up to 37 Mt announced by EU, carbon price plummeted from more than 30 €/t to 20 €/t.

The second appreciation period occurred in the summer of 2007. In this period, oil price played a key role in the rising carbon price. In fact, due to the shortages of global crude oil supply caused in part by tensions in Nigeria and limitations in the United States' ability to supply gasoline, oil price showed an upward trend which, in turn, pushed up carbon price, whereas carbon price was not sensitive to natural gas price reductions. Furthermore, in March 2007, the European Parliament announced that EU ETS would continue until 2020, which increased carbon futures price. Therefore, the observed and equilibrium values were all subjected to stimulated increase.

The second depreciation period occurred towards the end of Phase I when the carbon price fell due to its EUA being non-banking-EUA in Phase I was no longer suitable in Phase II. This decline had nothing to do with changes in underlying energy prices.

The third appreciation period occurred from January 2008 to August 2009, except for a downturn in the winter of 2009. Due to a declining carbon spot price, large amounts of carbon trading were transferred to carbon futures market. In addition, fine market expectation resulted in rising pressures on carbon price. In fact, in the first half of 2008, carbon futures market trading volumes accounted for more than 80% of the total trading volume. Moreover, the rise in oil price also played a key role in carbon price rise. During that period, with a shortage of global oil supply, oil price jumped to $147 per barrel, which was then its highest point in history. The rise in oil price pushed up carbon price. Then in October 2008, the onset of global financial crisis caused the depression in industrial production and the lack of market demand, which made the sharp decline in carbon price. At the beginning of 2009, carbon price fell below their equilibrium values. Then due to EU stimulations such as taking some effective measures, carbon market gradually recovered, and carbon price returned to, or exceeded, its equilibrium value.

The third depreciation period appeared after August 2009. With global financial crisis spreading to Europe, debt crisis was induced in Europe by the end of 2009. Some countries faced bankruptcy, industrial production activities stagnated and governments had no time to formulate, let alone implement, policies to address climate change. Moreover, with 2013 approaching and uncertain prospects looming in post-Kyoto times, carbon market expectations were gloomy, resulting in falling carbon price. At present, carbon price is still depreciating.

2.6 Conclusion

Based on the monthly EU ETS carbon futures price data, we investigated the driving factors behind carbon price using structure breakpoint tests, cointegration techniques, and ridge regression method. The following conclusions can be drawn.

First, 2007s Bali action plan, 2008s global financial crisis, and 2011s European debt crisis exerted significant influences over carbon price and caused the generation of structure breakpoints therein.

Second, a cointegration relationship existed between carbon price and its driving factors including energy prices, weather conditions, economic activities, and institutional decisions.

Third, equilibrium values showed that the observed carbon price had been lower than its equilibrium values since October 2009. Carbon price still tends to future depreciation.

References

Alberola E, Chevallier J, Cheze B (2008) Price drivers and structural breaks in European carbon prices 2005–2007. Energy Policy 36(2):787–797

Alberola E, Chevallier J, Cheze B (2009) European carbon price fundamentals in 2005–2007: the effects of energy markets, temperatures and sectorial production. J Policy Model 31(3): 446–462

Bai J, Perron P (2003) Computation and analysis of multiple structural change models. J Appl Econometrics 18:1–22

Benz E, Truck S (2009) Modeling the price dynamics of CO_2 emission allowances. Energy Econ 31(1):4–15

Bredin D, Muckley C (2011) An emerging equilibrium in the EU emissions trading scheme. Energy Econ 33:353–362

Chevallier J (2012) Econometric analysis of carbon markets. Springer, Berlin

Chevallier J, Ielpo F, Mercier L (2009) Risk aversion and institutional information disclosure on the European carbon market: a case-study of the 2006 compliance event. Energy Policy 37: 15–28

Christiansen A, Arvanitakis A, Tangen K, Hasselknippe H (2005) Price determinants in the EU emissions trading scheme. Climate Policy 5:15–30

Convery FJ, Redmond L (2007) Market and price developments in the European Union emissions trading scheme. Rev Environ Econ Policy 1(1):88–111

Creti A, Pierre-André J, Valérie M (2012) Carbon price drivers: Phase I versus Phase II equilibrium? Energy Econ 34:327–334

Drakakis K (2008) Empirical mode decomposition of financial data. Int Math Forum 3(25): 1191–1202

Feng ZH, Zou LL, Wei YM (2011) Carbon price volatility: evidence from EU ETS. Appl Energy 88:590–598

Gregory AW, Nason JM, Watt DG (1996) Testing for structural breaks in cointegrated relationships. J Econ 71:321–341

Guobrandsdottir HN, Haraldsson HO (2011) Predicting the price of EU ETS carbon credits. Syst Eng Procedia 1:481–489

Hintermann B (2010) Allowance price drivers in the first phase of the EU ETS. J Environ Econ Manag 59:43–56

Hoerl A, Kennard R (1970) Ridge regression: biased estimation for nonorthogonal problems. Technometrics 12(1):55–67

Kanen JLM (2006) Carbon trading and pricing. Environ Finance Publ, London

Keppler JH, Mansanet-Bataller M (2010) Causalities between CO_2, electricity, and other energy variables during phase I and phase II of the EU ETS. Energy Policy 38:3329–3341

Mansanet-Bataller M, Pardo AVE (2007) CO_2 prices, energy and weather. Energy J 28(3):73–92

Mansanet-Bataller M, Chevallier J, Herve-Mignucci M, Alberola E (2011) EUA and sCER phase II price drivers: unveiling the reasons for the existence of the EUA–sCER spread. Energy Policy 9:1056–1069

Oberndorfer U (2009) EU emission allowances and the stock market: evidence from the electricity industry. Ecology Econ 68(4):1116–1126

Paolella MS, Taschini L (2007) An econometric analysis of emission allowance prices. J Bank Finance 32(10):2022–2032

Seifert J, Uhrig-Homburg M, Wagner M (2008) Dynamic behavior of CO_2 spot prices. J Environ Econ Manag 56:180–194

The World Bank (2012) State and trends of the carbon market (2012). The World Bank, Washington, D.C.

Wei YM, Wang K, Feng ZH et al (2010) Carbon finance and carbon markets: method and empirical analysis. Science Press, Beijing

Zachmann G, von Hirschhausen C (2008) First evidence of asymmetric cost pass-through of EU emissions allowances: examining wholesale electricity prices in Germany. Econ Lett 99 (3):465–469

Zhang YJ, Wei YM (2010) An overview of current research on EU ETS: evidence from its operating mechanism and economic effect. Appl Energy 87(6):1804–1814

Chapter 3
Examining the Structural Changes of European Carbon Futures Price 2005–2012

Abstract This chapter deals with structural breaks detection techniques in the time series of carbon futures. The ICSS algorithm is detailed, followed by an event study analysis. Several explanations are advanced as to how and why the breakpoints occurred, related to high carbon era or, on the contrary, to the burst of the 2008 subprimes crisis and the 2011 European debt crisis.

3.1 Introduction

As an emerging financial market, the EU ETS has its own features. Compared with the financial markets, carbon price is more subjected not only to internal supplies and demands, but also to external heterogeneous environments such as intergovernmental negotiations, national allocation plans, and global economic or financial crisis, which make carbon price changes more complex. Further, structural changes of carbon prices may come into being, which have been a research hotspot (Zhang and Wei 2010). Thus, correctly examining the structural changes of European carbon price is beneficial not only in understanding the pricing mechanism of this emerging financial market, but also in effectively forecasting the future carbon price (Zhu and Wei 2013).

In recent years, more and more studies have investigated the structural changes of European carbon price. By establishing a multiple linear regression model with dummy variables, Alberola et al. (2008) found that institutional decision was one of the three notable factors of carbon price under the EU ETS. This finding was later re-proved by them in another work (Alberola et al. 2009). Chevallier et al. (2009) found that the certified data leakage event in May 2006 significantly impacted carbon price by using nonparametric kernel regression and generalized autoregressive conditional heteroskedasticity (GARCH) model. Creti et al. (2012) explored the drivers of carbon price by using the cointegration techniques. Their empirical results also show that the certified data leakage event in May 2006

Special thanks to Shujiao Ma and Yi-Ming Wei in assisting the writing of Chap. 3.

33

significantly affected carbon price. More recently, Zhu (2014) used an integrated model of Bai-Perron structure breakpoint test, cointegration technique, and ridge regression analysis to explore carbon price drivers. The empirical results show that the Bali action plan in 2007, global financial crisis in 2008 and European debt crisis in 2011 had significant impacts on carbon prices, which led to a structural breakpoint of carbon prices respectively.

The existing results can provide us with an important reference. However, there are also some shortcomings. First, they mainly employ the traditional econometric models; few use event study to estimate the effects of extreme events such as 2008 global financial crisis and 2011 European debt crisis on carbon price. Second, they mainly focus on carbon spot or over-the-counter (OTC) prices during the first stage (2005–2007), and only a few on carbon futures price during the second stage (2008–2012). However, carbon futures trading is the main stream, with the largest trading volume, which makes that carbon futures price may have more values (The World Bank 2012). Moreover, in theory, the sensitivity of carbon futures price to structural changes is lower than that of carbon spot price, which makes that the existing conclusions may not apply to carbon futures price. Therefore, it is hard to master the pricing mechanism of carbon market, and to effectively forecast carbon price. As a result, existing studies cannot provide decision-makers with sufficient information supports.

This chapter is designed to examine the structural changes of European carbon price over the two stages (2005–2012). Being different from the existed studies, this chapter employs the structural breakpoint test and event study models to explore the structural changes of European carbon price. Our results show that the extreme events: 2008 global financial crisis and 2011 European debt crisis generated three structural breakpoints in carbon price.

3.2 Methodology

3.2.1 Iterative Cumulative Sums of Squares (ICSS)

We use the ICSS algorithm, proposed by Inclan and Tiao (1994), to test the structural breakpoints in carbon price.

$p_t, t = 1, 2, \ldots, T$ is a carbon price series. $C_k = \sum_{t=1}^{k} p_t^2$, $k = 1, 2, \ldots, T$, is the cumulative sums of squares. $D_k = \frac{C_k}{C_t} - \frac{K}{T}$, $k = 1, 2, \ldots, T$, and $D_0 = D_T = 0$. $D^* = 1.358$ is chosen as the stopping condition.

Step 0: let $t_1 = 1$.

Step 1: calculate $D_k(a[t_1 : T])$ and let $K^*(\alpha[t_1 : T])$ be the corresponding k value of $\max_k |D_k(\alpha[t_1 : T])|$. Then find $M(t_1 : T) = \max_{t_1 \leq k \leq T}$ $\sqrt{(T - t_1 + 1)/2} |D_k(\alpha[t_1 : T])|$. If $M(t_1 : T) > D^*$, $K^*(\alpha[t_1 : T])$ is a breakpoint. Record this point and go to Step 2. If $M(t_1 : T) \leq D^*$, carbon price series cannot contain a breakpoint and the algorithm stops.

Step (2a): let $t_2 = K^*(\alpha[t_1 : T])$, and calculate $D_k(\alpha[t_1 : T]), t_1 = 1$.

    ① if $M(t_1 : t_2) > D^*$, there is a new possible breakpoint. Record this point and repeat Step (2a) until $M(t_1 : t_2) \leq D^*$.

    ② if $M(t_1 : t_2) \leq D^*$, there is no breakpoint in $t = t_1, \ldots, t_2$, the first breakpoint is $k_{first} = t_2$.

Step (2b): let $t_1 = k(\alpha[t_1 : t_2]) + 1$, and calculate $D_k(\alpha[t_1 : t_2])$, namely, calculate the D_k from the first breakpoint to the end of carbon price series.

    ① if $M(t_1 : t_2) > D^*$, there is a new possible breakpoint. Record this point and repeat Step (2b) until $M(t_1 : t_2) \leq D^*$.

    ② if $M(t_1 : t_2) \leq D^*$, there is no breakpoint in $t = t_1, \ldots, T$, the last breakpoint is $k_{last} = t_1 - 1$.

Step (2c): if $k_{first} = k_{last}$, there is only one breakpoint in the carbon price series, and the algorithm stops. If $k_{first} < k_{last}$, reserve these two possible breakpoints. Meanwhile, repeat Steps (1) and (2) on $\alpha[k_{first} = k_{last}]$. $\hat{N}_T$ is the number of breakpoints.

Step (3): if there are two or more breakpoints, construct the breakpoint vector $CP = (CP_0, \ldots, CP\hat{N}_{T+1})$ in ascending order. Then calculate the quantity $D_k(\alpha[CP_{j-1} + 1 : CP_{j+1}])$ when $j = 1, \ldots, \hat{N}_T$. If its value is greater than D^*, it can be determined that this point is a breakpoint. Otherwise, delete this point.

Step (4): repeat Steps (1)–(3) until the number of breakpoints is unchanged. Finally, all the structural breakpoints of carbon price are obtained.

3.2.2 Event Study

Further, we use the event study model (Demirer and Kutan 2010) to investigate the influence mode, intensity, and duration around the structural breakpoints of carbon price.

The return of carbon price is defined as formula (3.1):

$$R_t = \ln(p_t/p_{t-1}) \tag{3.1}$$

where R_t is the return of carbon price on day t, and p_t is carbon price on day t.

The expected or normal return is obtained by the market model as formula (3.2):

$$R_{t(\text{est.window})} = \alpha_{(\text{est.window})} + \beta_{(\text{est.window})} R_{mt(\text{est.window})} + \varepsilon_{t(\text{est.window})}$$

$$ER_{t(\text{est.window})} = \alpha_{(\text{est.window})} + \beta_{(\text{est.window})} R_{mt(\text{est.window})} ER_{t(\text{event.window})} \quad (3.2)$$

$$= \alpha_{(\text{est.window})} + \beta_{(\text{est.window})} R_{mt(\text{event.window})},$$

where $R_{t(\text{est.window})}$ is the return of carbon price on day t in the estimation window, $\alpha_{(\text{est.window})}$ is the intercept of carbon price in the estimation window, $\beta_{(\text{est.window})}$ is the slope of carbon price in the estimation window, $\alpha_{(\text{est.window})}$ and $\beta_{(\text{est.window})}$ are obtained by conducting ordinary least squares (OLS) regression analysis, $R_{mt(\text{est.window})}$ and $R_{mt(\text{event.window})}$ are the returns of the S&P composite index obtained from the Centre for Research on Securities Prices (CRSP) database on day t in the estimation and event windows, respectively. $ER_{t(\text{est.window})}$ and $ER_{t(\text{event.window})}$ are the expected return of carbon price on day t in the estimation and event windows, respectively.

The abnormal return is the difference between the actual return and normal return from a certain event on day t as formula (3.3):

$$AR_{t(\text{est.window})} = R_{t(\text{est.window})} - ER_{t(\text{est.window})}$$

$$AR_{t(\text{event.window})} = R_{t(\text{event.window})} - ER_{t(\text{event.window})} \quad (3.3)$$

where $AR_{t(\text{est.window})}$ and $AR_{t(\text{event.window})}$ are the abnormal return of carbon price on day t in the estimation and event windows, respectively. $R_{t(\text{est.window})}$ and $R_{t(\text{event.window})}$ are the return of carbon price on day t in the estimation and event windows, respectively.

Subsequently, the standard abnormal return (SAR) of carbon price on day t in the event window is as formulas (3.4) and (3.5):

$$SAR_t = \frac{AR_{t(\text{event.window})}}{\sqrt{S^2_{AR_t}}} \quad (3.4)$$

$$S^2_{AR_t} = \left[\frac{\sum_{t=a}^{b} \left[AR_{t(\text{est.window})} - \overline{AR}_{(\text{est.window})} \right]^2}{D-2} \right]$$

$$\times \left[1 + \frac{1}{D} + \frac{\left(R_{mt(\text{event.window})} - \overline{R}_{m(\text{est.window})} \right)^2}{\sum_{t=a}^{b} \left(R_{mt(\text{est.window})} - \overline{R}_{m(\text{est.window})} \right)^2} \right] \quad (3.5)$$

where $\overline{AR}_{(\text{est.window})}$ is the mean abnormal return of carbon price on day t in the estimation window, D is the trading day number in the estimation window, $\overline{R}_{m(\text{est.window})}$ is the mean return of the S&P composite index in the estimation window, and a and b are the starting and ending days of the estimation window, respectively.

To determine whether the standard abnormal return (SAR) is significantly equal to zero or not, a statistical measure is defined as (3.6)

$$z_t = \frac{\text{SAR}_t}{\sqrt{\frac{D-2}{D-4}}} \sim N(0,1) \tag{3.6}$$

The significance level of SAR_t, p, can be judged using the principle that z_t obeys a standard normal distribution. If $p < 0.05$, the result is significant at the 95% confidence level.

3.2.3   The ICSS-ES Model

To investigate and examine the structural changes of European carbon price, we propose an ICSS-ES model integrating the structural breakpoint test (ICSS) algorithm and event study (ES). The basic idea of the ICSS-ES model is that the ICSS algorithm is utilized to find out the structural breakpoints in carbon price series, and then event study is employed to verify the influence mode, intensity, and duration around the structural breakpoints of carbon price. The ICSS-ES model effectively integrates the ICSS algorithm, event study, and statistical analysis methods, which fully absorbs the advantages of these methods and can be effectively used to capture the structural changes of European carbon futures price. Its specific steps are as follows:

Step 1: Sort carbon futures price data, and extreme events affecting carbon market, including natural disasters, economic crises, institutional decisions, etc.

Step 2: Conduct structural breakpoint tests on carbon price data using ICSS to obtain the breakpoints.

Step 3: Compare the results of carbon price structural breakpoints with extreme events to investigate the relationship between carbon price changes and extreme events.

Step 4: Analyze the influence mode, intensity, and duration of around the structural breakpoints of carbon price using event study and statistical analysis methods.

3.3   Empirical Analysis

3.3.1   Data

In this chapter, we selected the following carbon price as the research sample: carbon futures price matured in December 2012, namely DEC12, from April 22, 2005 to April 30, 2012, obtained from the ECX, with 1800 daily trading prices in

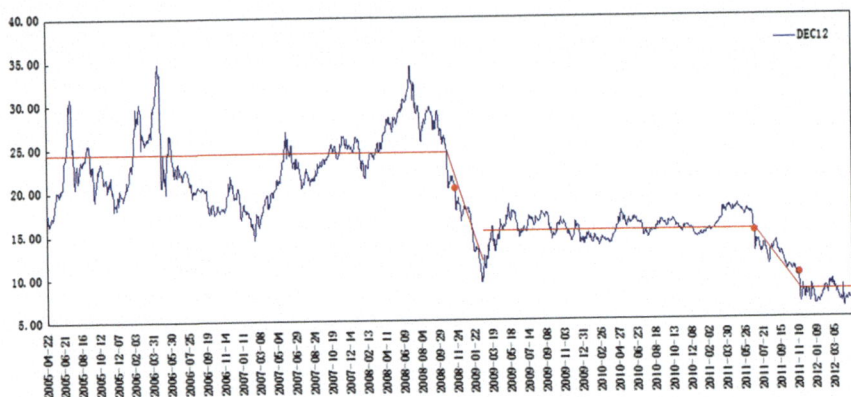

Fig. 3.1 Structural changes of DEC12 during 2005–2012

total. The reason that this contract is selected is that the purpose of this chapter is to empirically study the structural changes of European carbon futures price under the EU ETS during 2005–2012. Clearly, this contract is a representative contract throughout the two stages. Figure 3.1 presents the revolution of DEC12, with unit of euro/ton CO_2 equivalent.

3.3.2 Structural Breakpoint Test Using the ICSS Method

The ICSS algorithm has been employed to test the breakpoints in DEC12, and the obtained results are listed in Table 3.1. DEC12 contain 3 breakpoints. According to the dates of breakpoints, European carbon futures price can be divided into three steady and two slumping periods.

The first steady period (April 2005 to October 2008): In this period, carbon price was relatively stable, with a mean of 22.07 euro/ton. The EU ETS was in a "high carbon price era".

The first slumping period (November 2008 to February 2009): In September 2008, global financial crisis started. Economic panic gradually spread to many countries around the world. Industrial production was seriously affected and carbon market demand sharply declined. Consequently, carbon price slumped from 30 euro/ton to 10 euro/ton, and then gradually stayed at around 15 euro/ton.

The second steady period (March 2009 to May 2011): In this period, all countries around the world worked hand in hand. The world's major banks

Table 3.1 The breakpoints of DEC12

Carbon price	Breakpoint 1	Breakpoint 2	Breakpoint 3
DEC12	November 17, 2008	June 22, 2011	November 10, 2011

repeatedly cut interest rates and injected tremendous amounts of liquid capital into the markets. Moreover, many countries launched large-scale expansionary fiscal policies to stimulate economic growth. The global economy gradually recovered. During this period, carbon price was more stable and basically maintained around 15 euro/ton. This situation played a certain promoting role in helping some countries get out of the predicament they were in.

The second slumping period (June 2011 to October 2011): This slumping period originated from the Greek debt crisis, and showed some momentum to spread to core EU states such as France, etc., while other EU states such as Belgium, Spain, etc., were successively involved in the crisis. European economic recession led to carbon market fluctuations and reduced demand, which further caused a decline in carbon price. Carbon price slumped from above 15 euro/ton to below 10 euro/ton. Furthermore, the EU and International Monetary Fund (IMF) set about to solve the European debt crisis. They invested large amounts of aiding funds and launched massive policies to stimulate the economy. By doing this, carbon price tended to become steady and fluctuated at around 8 euro/ton.

The third steady period (November 2011 to present day). The European debt crisis worsened again. EU states have become successively involved in the debt crisis. Germany auctioned debt, while the ratings of Portugal, Hungary, etc., were downgraded in sequence. This trigged economic recession in European countries, which caused further decline in carbon price. Carbon price is, at present, generally around the low level of 7 euro/ton.

In summary, each surge and slump in carbon price is resulted from an extreme event. The first slump originated from 2008 global financial crisis, while the second slump is related to 2011 European debt crisis.

3.3.3 Structural Changes Analysis Using the ES Model

First, we examine the first breakpoint of DEC12. To investigate the structural changes using the ES model, the estimation window and the event window should first be determined. We take period 0 to represent November 17, 2008, so that negative numbers represent the days before, and positive ones stand for the days afterwards. Ten days before and after this event are selected (from −10 to 10) as the event window. 100 days before the event are chosen as the estimation window. Thus, the range of the estimation window refers to the period from the 110th day to the 11th day before the event (from −110 to −11). Figure 3.2 illustrates the timing of the event study.

The results are shown in Table 3.2 and Fig. 3.3. DEC12 presents the maximum return value on the third day after the breakpoint. Meanwhile, the p-value is 0.0003, which is smaller than 0.01. Thus, the result is significant at the 1% level. Moreover, in the event window 10 days after the first breakpoint in DEC12, the p-values of three days are lower than 0.05. This suggests that the return fluctuations are actually

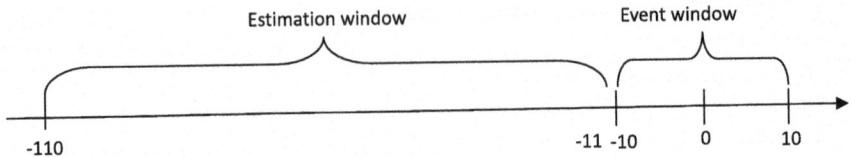

Fig. 3.2 Event study timeline

Table 3.2 Abnormal returns at the first breakpoint of DEC12

Date	Abnormal return	Standard abnormal return	z-statistic	p-value
−10	0.0176	0.7568	0.7490	0.4538
−9	0.0385	1.6575	1.6405	0.1009
−8	−0.0111	−0.4764	−0.4715	0.6373
−7	−0.0113	−0.4877	−0.4827	0.6293
−6	0.0347	1.4934	1.4781	0.1394
−5	0.0041	0.1753	0.1735	0.8623
−4	−0.0186	−0.7993	−0.7911	0.4289
−3	−0.0052	−0.2224	−0.2201	0.8258
−2	−0.0001	−0.0040	−0.0040	0.9968
−1	−0.0135	−0.5796	−0.5737	0.5662
0	−0.0084	−0.3593	−0.3556	0.7221
1	−0.0159	−0.6858	−0.6788	0.4973
2	−0.0082	−0.3548	−0.3512	0.7254
3	−0.0859	−3.6918	−3.6540	0.0003
4	0.0110	0.4742	0.4693	0.6388
5	0.0552	2.3735	2.3491	0.0188
6	0.0024	0.1043	0.1033	0.9178
7	−0.0095	−0.4105	−0.4063	0.6846
8	0.0475	2.0444	2.0235	0.0430
9	−0.0329	−1.4167	−1.4022	0.1609
10	0.0003	0.0108	0.0107	0.9915

big near this breakpoint. Meanwhile, it verifies that 2008 global financial crisis had a great influence on carbon price.

Second, we examine the second breakpoint of DEC12. Similar to the event study in the first breakpoint, and the results are shown in Table 3.3 and Fig. 3.4. In the event window of 20 days before and after the second breakpoint in DEC12, 10 days are significant at the 5% level. This means that there are certainly return fluctuations near the second breakpoint, and so carbon price shows obvious and abnormal fluctuations.

In January 2011, Fitch downgraded Greece's credit rating from BBB− to BB+, and the rating outlook was rated as negative. However, obvious fluctuation was not observed in the EU ETS and carbon price remained at about 15 euro/ton. Shortly

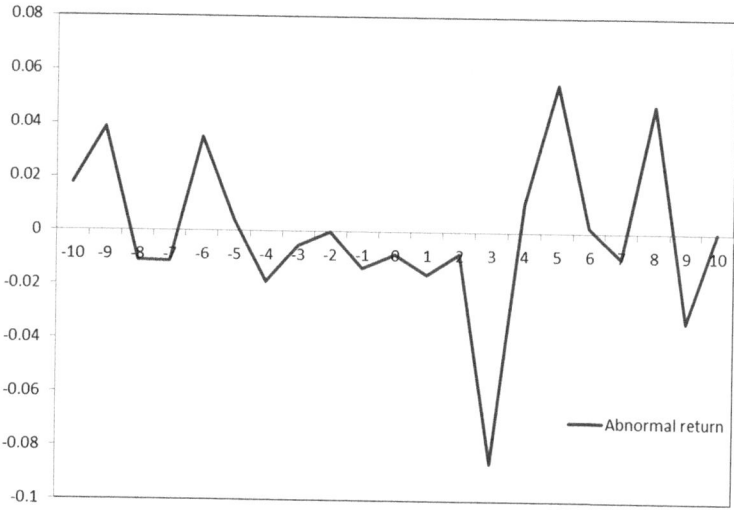

Fig. 3.3 Abnormal returns at the first breakpoint of DEC12

Table 3.3 Abnormal returns at the second breakpoint of DEC12

Date	Abnormal return	Standard abnormal return	z-statistic	p-value
−10	−0.0097	−0.7451	−0.7375	0.4608
−9	0.0093	0.7124	0.7051	0.4808
−8	−0.0028	−0.2182	−0.2160	0.8290
−7	0.0006	0.0465	0.0460	0.9633
−6	−0.0114	−0.8795	−0.8705	0.3840
−5	−0.0173	−1.3337	−1.3201	0.1868
−4	−0.0104	−0.8032	−0.7950	0.4266
−3	−0.0292	−2.2479	−2.2248	0.0261
−2	−0.0262	−2.0174	−1.9967	0.0459
−1	−0.0314	−2.4122	−2.3874	0.0170
0	−0.0022	−0.1688	−0.1671	0.8673
1	−0.1006	−7.7300	−7.6507	0.0000
2	−0.0902	−6.9206	−6.8497	0.0000
3	0.0592	4.5567	4.5100	0.0000
4	0.0435	3.3453	3.3110	0.0009
5	−0.0370	−2.8464	−2.8172	0.0048
6	0.0382	2.9410	2.9108	0.0036
7	−0.0203	−1.5617	−1.5457	0.1222
8	0.0096	0.7352	0.7277	0.4668
9	0.0101	0.7769	0.7690	0.4419
10	−0.0281	−2.1580	−2.1358	0.0327

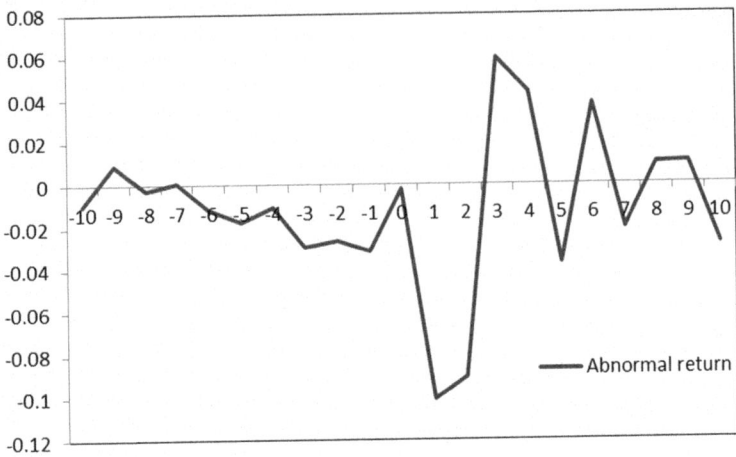

Fig. 3.4 Abnormal returns at the second breakpoint of DEC12

afterwards, on March 7, 2011, Moody's downgraded Greece's credit rating from BA1 to B1, and the rating outlook was rated as negative. Standard & Poor's downgraded Greece's credit rating from BB+ to BB−. Then, the EU ETS began to fall into the earlier stage of crisis. In June 2011, the Greek parliament approved a 5-year financial retrenchment plan. Unfortunately, crisis spread had not been eased. In July, Standard & Poor's downgraded Greece's long-term credit rating from B to CCC. Also, it pointed out that the debt rollover plan could put Greece into a selective default position. This would take Greece's crisis to a deeper level. Since then, the debt crisis has begun to spread across Europe, and carbon price declined from the sustained level of 15 euro/ton to below 10 euro/ton. Subsequently, to prevent spreading of the crisis, European countries launched a series of aid measures. However, these measures have not managed to control crisis spread effectively. Subsequently, other countries have fallen into the crisis successively and the crisis has even begun to encumber the whole Eurozone.

Finally, we examine the third breakpoint of DEC12. Similar to the event study in the former two breakpoints, and the results are shown in Table 3.4 and Fig. 3.5. In the event window of 10 days after the third breakpoint in DEC12, 6 days are significant at the 5% level. Abnormal returns reach the maximum value on the first day after the breakpoint. The p-value is 0, which shows that there is abnormal return around this breakpoint. Therefore, it can be deduced that this event really had a significant influence on carbon price.

Since November 2011, European countries have been trapped in the crisis spreading panic. At the beginning of the month, Greece's prime minister and president announced their resignations. Mid-month, Italy's prime minister announced his resignation due to the crisis. Then, the Spanish national debt return exceeded the warning line of 7%. Spain pulled down the debt crisis alarm. The election was brought forward and Prime Minister Zapatero's government stepped

Table 3.4 Abnormal returns at the third breakpoint of DEC12

Date	Abnormal return	Standard abnormal return	z-statistic	p-value
−10	0.0330	1.3275	1.3139	0.1889
−9	0.0026	0.1054	0.1043	0.9169
−8	−0.0187	−0.7520	−0.7443	0.4567
−7	−0.0243	−0.9783	−0.9683	0.3329
−6	−0.0316	−1.2710	−1.2580	0.2084
−5	0.0221	0.8898	0.8807	0.3785
−4	−0.0250	−1.0056	−0.9953	0.3196
−3	0.0569	2.2906	2.2671	0.0234
−2	0.0161	0.6470	0.6404	0.5219
−1	−0.0171	−0.6900	−0.6830	0.4946
0	−0.0017	−0.0703	−0.0696	0.9445
1	−0.2403	−9.6585	−9.5594	0.0000
2	−0.0571	−2.2979	−2.2743	0.0229
3	−0.0713	−2.8711	−2.8416	0.0045
4	0.0522	2.1025	2.0809	0.0374
5	−0.0022	−0.0904	−0.0895	0.9287
6	0.0362	1.4594	1.4445	0.1486
7	0.1869	7.5127	7.4356	0.0000
8	−0.0622	2.5030	−2.4773	0.0132
9	−0.0297	−1.1975	−1.1853	0.2359
10	0.0207	0.8336	0.8250	0.4094

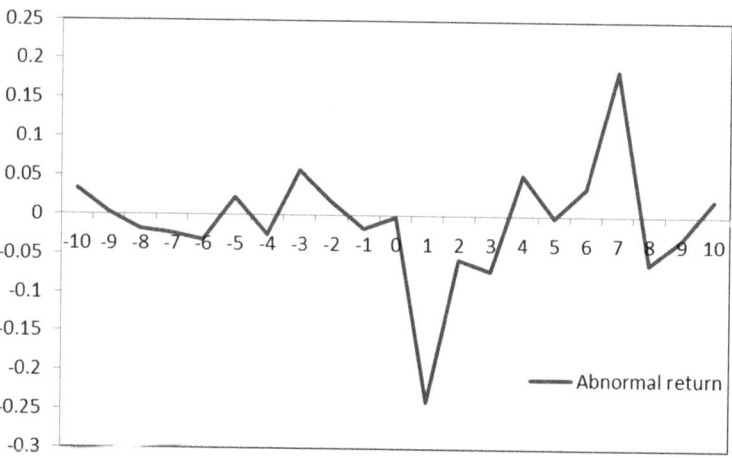

Fig. 3.5 Abnormal returns at the third breakpoint of DEC12

down. Up to then, the crisis had not weakened but instead had become worse. At the end of November, Portugal's credit debt rating was rated as junk grade. The Hungarian foreign currency bond was rated as BA1. Subsequently, Moody's credit rating agency stated that European debt crisis was threatening the credit conditions of all the European countries. This implied that even the countries with AAA grade rating, namely, Germany, France, Austria and Holland were all threatened. Since November 2011, carbon price has dropped from 10 to 6 euro/ton, and stayed at a low level of about 7 euro/ton.

In summary, the breakpoints in DEC12 obtained from the ICSS algorithm all happened in the breakout period of 2008 global financial crisis and 2011 European debt crisis. DEC12 has three breakpoints. The first one is in November 2008, which is related to the 2008 global financial crisis. The second and the third are in June and November 2011, respectively, which are related to the European debt crisis deterioration and market panic periods. By event study, we specifically estimated the event windows 20 days before and after each breakpoint to explore the influence mode, intensity, and duration of around the structural breakpoints of carbon price. The results suggest that, at the 5% significance level, 2008 global financial crisis and 2011 European debt crisis both had significant impacts on carbon price. Based on this, we can say with confidence of above 95% that the two extreme events can cause large fluctuations in carbon price, and three structural breakpoints in carbon price.

3.4 Conclusion

Based on the daily European futures price, we investigate the structural changes of European carbon futures prices using the structure breakpoint test: iterative cumulative sums of squares algorithm and event study models. The following conclusions can be drawn: Since 2005, there have been three breakpoints of European carbon futures price, arisen from the two extreme events of 2008 global financial crisis and 2011 European debt crisis. Furthermore, due to the great effects of extreme events on carbon price, more attention should be paid to the effects of extreme events on carbon price in future forecasting processes. By building on the knowledge gained from this chapter and setting up different scenarios, carbon price prediction can be modified to obtain more flexible and adaptive results.

References

Alberola E, Chevallier J, Cheze B (2008) Price drivers and structural breaks in European carbon prices 2005–2007. Energy Policy 36(2):787–797
Alberola E, Chevallier J, Cheze B (2009) European carbon price fundamentals in 2005–2007: the effects of energy markets, temperatures and sectorial production. Policy Model 31(3):446–462

Chevallier J, Ielpo F, Mercier L (2009) Risk aversion and institutional information disclosure on the European carbon market: a case-study of the 2006 compliance event. Energy Policy 37:15–28

Creti A, Jouvet PA, Mignon V (2012) Carbon price drivers: Phase I versus Phase II equilibrium? Energy Econ 34:327–334

Demirer R, Kutan AM (2010) The behavior of crude oil spot and futures prices around OPEC and SPR announcements: an event study perspective. Energy Econ 32:1467–1476

Inclan C, Tiao GC (1994) Use of cumulative sums of squares for retrospective detection of change of variance. J Am Stat Assoc 89(427):913–923

The World Bank (2012) State and trends of the carbon market. The World Bank

Zhang YJ, Wei YM (2010) An overview of current research on EU ETS: evidence from its operating mechanism and economic effect. Appl Energy 87(6):1804–1814

Zhu BZ (2014) Carbon price drivers: evidence from EU ETS. J Beijing Inst Technol (Soc Sci Ed) 16(3):10–16

Zhu BZ, Wei YM (2013) Carbon price prediction with a hybrid ARIMA and least squares support vector machines methodology. Omega 41(3):517–524

Chapter 4
A Multiscale Analysis for Carbon Price with Ensemble Empirical Mode Decomposition

Abstract This chapter develops ensemble empirical mode decomposition and fine-to-coarse reconstruction in order to extract carbon price signals from a multiscale viewpoint. The decomposition shows the carbon price is affected by both long-term (e.g., trend) and short-term (e.g., supply–demand fundamentals) imbalances that require appropriate forecasting strategies.

4.1 Introduction

The origin of the global carbon market date back to 1992 when 180 countries signed the United Nations Framework Convention on Climate Change (UNFCCC) which had the stated goal to "stabilize atmospheric greenhouse gas (GHGs) concentrations at 'safe' levels." Following negotiations under this agreement, the Kyoto Protocol was signed in December 1997, committing the industrialized nations to an averaged 5.2% reduction from 1990 levels by the first committing period from 2008 to 2012. With ratification by the European nations, Canada, Japan, New Zealand, and now Russia, the Kyoto Protocol went into effect on February 16, 2005—an indication that governments have begun to get moving on mitigating global warming by reducing GHGs emissions. Under the Kyoto Protocol, different economies agree to take common but differentiated obligations, and the European Union (EU) has proved to be a frontrunner in implementing the emission reduction targets specified by the Kyoto Protocol and addressing climate changes. According to the Kyoto Protocol, the EU is committed to reducing GHGs emissions by 8% below its 1990 level during the period from 2008 to 2012.

In order to meet the commitment of the Kyoto protocol at the lowest overall cost or most economically efficient solution, also in 2005, the EU ETS for GHGs was launched within the EU covering around 12,000 installations in 25 countries and 6 major industrial sectors. This new system is a drastic change of market paradigm, and

Special thanks to Ping Wang, Dong Han, and Ying-Ming Wei in providing research assistance for Chap. 4.

© Springer International Publishing AG 2017
B. Zhu and J. Chevallier, *Pricing and Forecasting Carbon Markets*,
DOI 10.1007/978-3-319-57618-3_4

the first step of EU member states to implement emission reductions agreed in the Kyoto Protocol, as well as the first multinational GHGs cap-and-trade system in the world. In terms of market value or trading volume whatever, the EU ETS is the largest carbon market all over the world up to date. The market value and trading volume of EU ETS are much higher than that of New South Wales Greenhouse Gas Abatement Scheme (GGAS), Chicago Climate Exchange (CCX), and other main international carbon markets, and also significantly exceed that of project-based carbon markets including Clean Development Mechanism (CDM). As the vane of international carbon market, the development of EU ETS reflects the direction of global carbon market and the price of EU ETS directly influences the reference price of global carbon market (Zhang and Wei 2010). Therefore, it is of great theoretical and practical significance to a multiscale analysis for the EU ETS carbon price from a novel perspective.

Carbon markets especially the EU ETS has proven to be not only an important tool for human beings to address climate changes, but also a major choice for investors to decentralize their investment risks. In recent years, as a result of international carbon price fluctuation, the volatility of carbon price has presented such high complex characteristics as nonlinear and nonstationary, which makes the carbon price analysis and prediction to be one of the key issues concerned by many academic researchers and business practitioners. In this context, this chapter carries out a multiscale analysis on the period, strength, characteristics and causes of international carbon price fluctuation, extracts its feature timescales, reveals its periodic characteristics, and explores the main factors affecting carbon price fluctuation. They are aimed to deeply understand the EU ETS carbon price formation mechanism, grasp the rules of carbon price fluctuation, and predict the future carbon price trends, which can not only help governments formulate correct carbon price policies to guarantee the safe operation of carbon market, but also help investors take effective measures to mitigate the shock of the carbon price volatility and eliminate the negative impact caused by carbon price fluctuation so as to evade their investment risks associated with carbon market.

During the past few years, there have been some studies on analysis and forecasting of the EU ETS carbon price. The approaches applied can be grouped into two categories: structure models and data-driven methods. The standard structure models outline the EU ETS carbon market and then analyze the carbon price volatility mainly from the perspective of supply–demand equilibrium of carbon market (Kanen 2006; Reilly and Paltsev 2005; Seifert et al. 2008). Data-driven models include linear models such as multivariate regression (MR), autoregressive moving average (ARMA), vector autoregressive(VAR), Granger causality test (GCT), and autoregressive conditional heteroscedasticity (ARCH) type models including ARCH, generalized ARCH (GARCH), threshold GARCH (TGARCH), and Markov switching (MS)—autoregressive(AR)-GARCH (Zhang and Wei 2010; Bataller et al. 2007; Paolella and Taschini 2006; Beat 2010; Bunn and Fezzi 2008; Keppler and Bataller 2010), and nonlinear models such as differential equation (DE) models (Montagnoli and de Vries 2009) and rescaled range (R/S) analysis model (Feng et al. 2011). A number of other references on this topic exist (Springer 2003; Christiansen et al. 2005; Paolella and Taschini 2006; Alberola et al. 2008a, b, 2009; Chevallier et al. 2009; Chevallier 2009).

To sum up, the existing literatures can provide us with important references, but empirical studies on carbon market price formation mechanism have just been started up and they still have some shortages. The structure models can help understand the mechanisms of carbon price determination and quantify each factor's impact on carbon price. However, this approach has proved to be difficult to be implemented in practice due to some specific characteristics of the EU ETS carbon market. For example, it is hard to model the supply because carbon emission allowance is supplied by many independent owners including government, banks, investors, etc.; meanwhile, the dynamic and unstable market environment increases the difficulty of modeling. Data-driven methods often perform well when applied to short-term forecasting but they lack economic meaning and cannot well explain the inner driving forces that move the EU ETS carbon price.

Facing with the dilemma between difficulties in modeling and lack of economic meaning, Huang et al. (1998) from National Aeronautics and Space Administration (NASA) put forward a new data analysis method, i.e., empirical mode decomposition (EMD), which is well positioned to solve such problems. EMD is an empirical, intuitive, direct, and self-adaptive data processing method which is proposed especially for nonlinear and nonstationary data. The core of EMD is to decompose data into a smaller number of independent and nearly periodic intrinsic modes based on local characteristic scale. Each derived intrinsic mode defined as the distance between two successive local extrema in EMD is dominated by scales in a narrow range. Thus, according to the scale, the concrete implications of each mode can be identified. For example, an intrinsic mode derived from an economic time series with a scale of 3 months can often be recognized as the seasonal component. Since carbon price data is the only link we have with the reality, by exploring data's intrinsic modes, EMD not only helps discover the characteristics of the data but also helps understand the underlying rules of reality. Compared with other multiscale analyses including wavelet analysis, EMD algorithm can reflect the physical properties of the observed data more accurately and has stronger local performance capacity. Therefore, EMD is more effective in dealing with the processing of nonlinear and nonstationary carbon price data.

EMD was initially proposed for study of ocean waves, and then successfully applied in many areas, such as biomedical engineering, structured health monitoring, earthquake engineering, and global primary productivity evolution. However, these applications are mainly limited to studies of nature science and engineering. In recent years, there have been only a few successful applications in social sciences so far. The first is by Huang et al. (2003), to apply EMD to financial data to examine the changeability of the markets. The second is by Cummings et al. (2004), to use EMD to prove the existence of a spatial–temporal traveling wave in the incidence of dengue haemorrhagic fever in Thailand. In addition, EMD has been successfully used for West Texas Intermediate (WTI) crude oil spot price analysis (Zhang et al. 2008) and financial crisis forecasting (Yu et al. 2010). These empirical results showed that EMD can be more widely used in social sciences.

By way of EMD algorithm, carbon price can be decomposed into several nearly independent intrinsic modes, which have different timescales and represent different carbon price volatilities caused by different factors, thus simplifying the interference and coupling across characteristic information of different scales in carbon price series. Therefore, this chapter applies ensemble empirical mode decomposition (EEMD), an improved version of EMD, to EU ETS carbon future price data and finds that it can help interpret the formation of EU ETS carbon price from a novel perspective. First, three carbon price series with different scales are decomposed into several independent IMFs and a residue from high frequency to low frequency. Second, the intrinsic modes and residue are composed into a high-frequency component, a low-frequency component, and a trend component based on fine-to-coarse reconstruction. Depending on the scales and characteristics of these three components, their economic meanings are identified as short-term fluctuations caused by normal supply–demand disequilibrium or some other market activities, the effect or shock of a significant event, and a long-term trend, respectively. Finally, by defining the features of these three components, some forecasting strategies for EU ETS carbon price are also discussed at the end of this chapter.

4.2   Methodology

4.2.1   Empirical Mode Decomposition

EMD is generally a nonlinear and nonstationary data processing method developed by Huang et al. (1998) and Huang et al. (1999). It assumes that the EU ETS carbon price, depending on its complexity, may simultaneously have many different coexisting modes of oscillations. EMD can extract these intrinsic modes from the original EU ETS carbon price time series based on the local characteristic scale of the EU ETS carbon price data themselves, and represent each intrinsic mode as an intrinsic mode function (IMF), which meets the following two conditions:

(1) The functions have the same number of extrema and zero crossings or differ by one at the most;
(2) The functions are symmetric with the local zero mean.

The two conditions ensure that an IMF is a nearly periodic function and the mean is set to zero. IMF is a harmonic-like function, but with variable amplitude and frequency at different times. In practice, the IMFs are extracted through a sifting process. The EMD algorithm is described as follows:

(1) Identify all the maxima and minima of carbon price time series $x(t)$;
(2) Generate their upper and lower envelopes, $e_{max}(t)$ and $e_{min}(t)$, with cubic spline interpolation.
(3) Calculate the point-by-point mean $m(t)$ from the upper and lower envelopes as formula (4.1):

$$m(t) = \frac{e_{\max}(t) + e_{\min}(t)}{2}. \tag{4.1}$$

(4) Extract the mean from carbon price time series and define the difference between $x(t)$ and $m(t)$ as $d(t)$:

$$d(t) = x(t) - m(t). \tag{4.2}$$

(5) Check the properties of $d(t)$:

    ① If it is an IMF, denote $d(t)$ as the ith IMF and replace $x(t)$ with the residue $r(t) = x(t) - d(t)$. The ith IMF is often denoted as $c_i(t)$ and the i is called its index;

    ② If it is not an IMF, replace $x(t)$ with $d(t)$;

(6) Repeat steps (1)–(5) until the residue satisfies some stopping criteria.

According to the prerequisites mentioned above, one typical stopping criterion proposed by Huang et al. (1998) for extracting an IMF is the residue $d(t)$ satisfies the following stopping condition (4.3):

$$SD = \sum_{t=2}^{T} \frac{|d_{k-1}(t) - d_k(t)|^2}{d_{k-1}^2(t)}, \tag{4.3}$$

where $d_{k-1}(t)$ is the sifting result in the $k-1$th iteration, and SD is the stopping condition. Typically, SD is usually set between 0.2 and 0.3.

As an improvement to the criteria proposed by Huang et al. (1998) and Huang et al. (1999) that have been considered so far, Rilling et al. (2003), a well-known French scholar, introduced a new criterion based on two thresholds θ_1 and θ_2, aimed to guarantee globally small fluctuations in the mean while taking into account locally large excursions. This amounts to introduce the mode amplitude $\alpha(t)$ as (4.4)

$$\alpha(t) = \frac{|e_{\max}(t) + e_{\min}(t)|}{2}, \tag{4.4}$$

and the evaluation function $\sigma(t)$ as (4.5)

$$\sigma(t) = \left|\frac{m(t)}{\alpha(t)}\right|, \tag{4.5}$$

so that sifting is iterated until $\sigma(t) < \theta_1$ for some prescribed fractions $(1 - \alpha)$ of the total duration, while $\sigma(t) < \theta_2$ for the remaining fractions. One can typically set $\alpha = 0.05$, $\theta_1 = 0.05$, and $\theta_2 = 10\theta_1$.

The EMD extracts the next IMF by applying the sifting procedure above to the residue term $r_m(t) = x(t) - c_1(t)$, where $c_1(t)$ denotes the first IMF. The decomposition process can be repeated until the last residue $r_m(t)$ has at most one local extremum or becomes a monotonic function, from which no more IMFs can be extracted. The sifting procedure above can be implemented using Matlab software.

At the end of this sifting procedure, carbon price data series $x(t)$ can be expressed as formula (4.6):

$$x(t) = \sum_{i=1}^{m} c_i(t) + r_m(t), \tag{4.6}$$

where m is the number of IMFs, $r_m(t)$ is the final residue, which is the main trend of $x(t)$, and $c_i(t)$, $(i = 1, 2, \ldots, m)$ are the IMFs, which are nearly orthogonal to each other, and all have nearly zero means. Thus, one can achieve the decomposition of carbon price data series into m IMFs and one residue. The IMF components contained in each frequency band are different and they change with variation of carbon price $x(t)$, while $r_m(t)$ represents the central tendency of carbon price $x(t)$.

Compared with the traditional Fourier and wavelet decompositions, the EMD technique has several distinct advantages. First of all, EMD is relatively easy to be understood and implemented. Second, since the decomposition is based on the local characteristic timescale of the data and only extrema are used in the sifting process, EMD is local, self-adaptive, concretely implicational and highly efficient for nonlinear and nonstationary time series decomposition. Hence, EMD can adaptively and robustly decompose carbon price time series into several independent IMF components and one residue component. IMFs and the residue component displaying linear and nonlinear behaviors depend only on the nature of carbon price time series being studied. Third, the IMFs derived from EMD decomposition have a clear instantaneous frequency as the derivative of the phase function, so Hilbert transformation can be applied to the IMFs, allowing us to analyze the data in a time–frequency–energy space. Last but not the least, in wavelet decomposition, a filter base function must be determined beforehand, but it is difficult for some unknown series to determine the filter base function. Unlike wavelet decomposition, EMD is not required to determine a filter base function before decomposition. In terms of the merits above, the EMD can be used as an effective decomposition tool.

4.2.2   Ensemble Empirical Mode Decomposition

EMD has proved to be quite versatile in a broad range of applications for extracting signals from data generated in nonlinear and nonstationary processes. However, the original EMD has a drawback—the frequent appearance of mode mixing, which is defined as a single IMF either consisting of signals of widely disparate scales, or a signal of a similar scale residing in different IMF components. To overcome the problem, Wu and Huang (2009) proposed EEMD. The basic idea of EEMD is that each observed data is amalgamations of the true time series and noise. Thus, even if data are collected by separate observations, each with a different noise level, the ensemble mean is close to the true time series. Therefore, an additional step is taken by adding white noise that may help extract the true signal in the data. The procedure of EEMD is developed as follows:

(1) Add a white noise series to the targeted data;
(2) Decompose the data with added white noise into IMFs;
(3) Repeat step (1) and step (2) iteratively, but with different white noise series each time, and obtain the (ensemble) means of corresponding IMFs of the decompositions as the final result.

The added white noise series present a uniform reference frame in the time–frequency and timescale space for signals of comparable scales to collate in one IMF and then cancel itself out (via ensemble averaging), after serving its purpose; therefore, it significantly reduces the chance of mode mixing and represents a substantial improvement over the original EMD. The effect of the added white noise can be controlled according to the well-established statistical rule proved by Wu and Huang (2009) as (4.7)

$$\varepsilon_n = \frac{\varepsilon}{\sqrt{n}}, \tag{4.7}$$

where n is the number of ensemble members, ε is the amplitude of the added noise, and ε_n is the final standard deviation of error, which is defined as the difference between the input signal and the corresponding IMFs. In practice, the number of ensemble members is often set to 100 and the standard deviation of white noise series is set to 0.1 or 0.2.

4.2.3   Fine-to-Coarse Reconstruction

In the sifting process of EEMD, the first component, c_1, contains the finest scale (or the shortest period component) of carbon price time series. The residue after extracting c_1 contains variations for a longer period in the data. Therefore, the modes are extracted from high frequency to low frequency. EEMD can be used as a filter to separate high-frequency (fluctuating process) and low-frequency (slowly varying component) modes. In practice, the following algorithm, based on fine-to-coarse reconstruction, i.e., high-pass filtering by adding fast oscillations (IMFs with smaller index) up to slow (IMFs with larger index) is adopted:

(1) Compute the mean of the sum of c_1 to $c_i(1 \leq i \leq m)$, i.e., $s_i = \sum_{k=1}^{i} c_k$ for each component (except for the residue);

(2) Select the significance level α and employ t test to identify for which i the mean significantly departs from zero for the first time;

(3) Once i is identified as a significant change point, partial reconstruction with IMFs from this to the end is identified as the slowly varying mode, i.e., low-frequency component, and the partial reconstruction with other IMFs is identified as the fluctuating process, i.e., high-frequency component. In addition, the residue is identified as the trend mode, i.e., trend component.

4.3 Decomposition

Through EEMD, carbon price can be decomposed into a set of independent IMFs with different timescales, and one residue. The analyses of these IMFs and the residue can help explore the variability and formation of EU ETS carbon price from a new perspective.

4.3.1 Data

The chosen daily data of ECX carbon future price is from April 22, 2005 to January 12, 2016, 2747 trading prices excluding public holidays in total. Inspired by Koenig (2011), the daily settlement prices of EUA futures at three phases are combined as the "EUA Tracker" of carbon market in this research, including the settlement prices from April 22, 2005 to December 15, 2008 (DEC08), those from December 16, 2008 to December 17, 2012 (DEC12) and those from December 18, 2012 to January 12, 2016 (DEC16). The first two EUA futures matured in December 2008 and 2012, and the last one will mature in December 2016. Figure 4.1 shows carbon price data series of ECX in unit of euros/ton carbon dioxide. The reason for EUA futures is that first, as the largest carbon market of the EU ETS, ECX accounts for over 70% of carbon futures everyday in the world. Second, compared with EUA spot goods, EUA futures present a larger trading volume. Third, the futures price has the ability of price discovery (Uhrig-Homburg and Wagner 2009; Rittler 2012). Figure 4.1 shows that carbon price moves in a relatively stable trend as a whole, besides nearly periodic fluctuations of different scales and other abnormal fluctuations. As a result, there are many difficulties in accurately analyzing and forecasting complex carbon price.

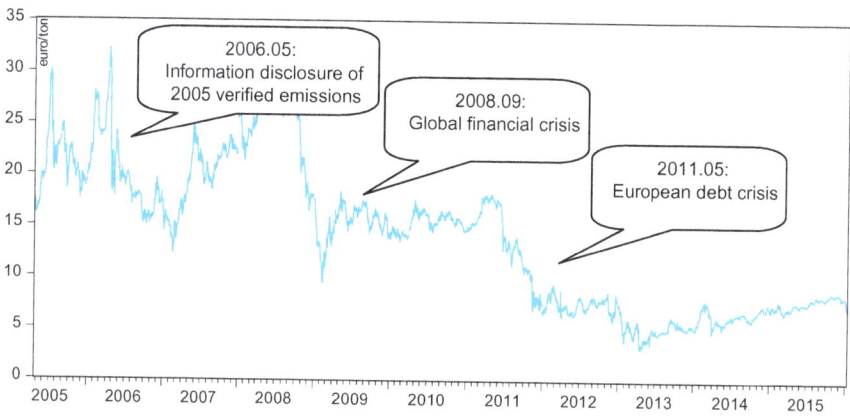

Fig. 4.1 ECX carbon price from April 22, 2005 to January 12, 2016

4.3.2   IMFs

The above-mentioned carbon price data set is decomposed into several independent IMFs and the residue derived from applying EEMD as shown in Fig. 4.2. During the process of EEMD, the number of ensemble members is set to 100 and the standard deviation of white noise series is set to 0.1, besides that, the number of sifting iteration is set to 10. Since the number of IMFs will be restricted to $\log_2 N$, where N is the number of samples, the sifting processes produce ten IMFs plus one residue for daily data. All the IMFs are listed in the order in which they are extracted, that is, from the highest frequency to the lowest frequency. The last item in the figure is the residue. Compared with the original carbon price data set, EEMD decomposition does not produce significant mode overlapping and endpoint effect, which shows its effectiveness. Analyses of such IMFs and residue can help find out the volatility characteristics and formation mechanism.

All the IMFs present as changing frequencies and amplitudes, unlike any harmonic. The frequencies and amplitudes of all the IMFs vary over time. With the frequency changing from high to low, the amplitudes of the IMFs are becoming larger: for example, the amplitudes of IMF1 in Fig. 4.2 are smaller than 2, but the amplitudes of IMF8 in Fig. 4.2 are restricted to nearly 8. The last residue is mode slowly varying around the long-term average. Corresponding to the original carbon price time series, the IMFs were sharply shocked during the trial period of 2005–2007, the financial crisis of 2008/2009 and the European debt crisis of 2010–2011, but amplitude fluctuations rapidly decreased and got narrow after 2012 for a relative stable external condition.

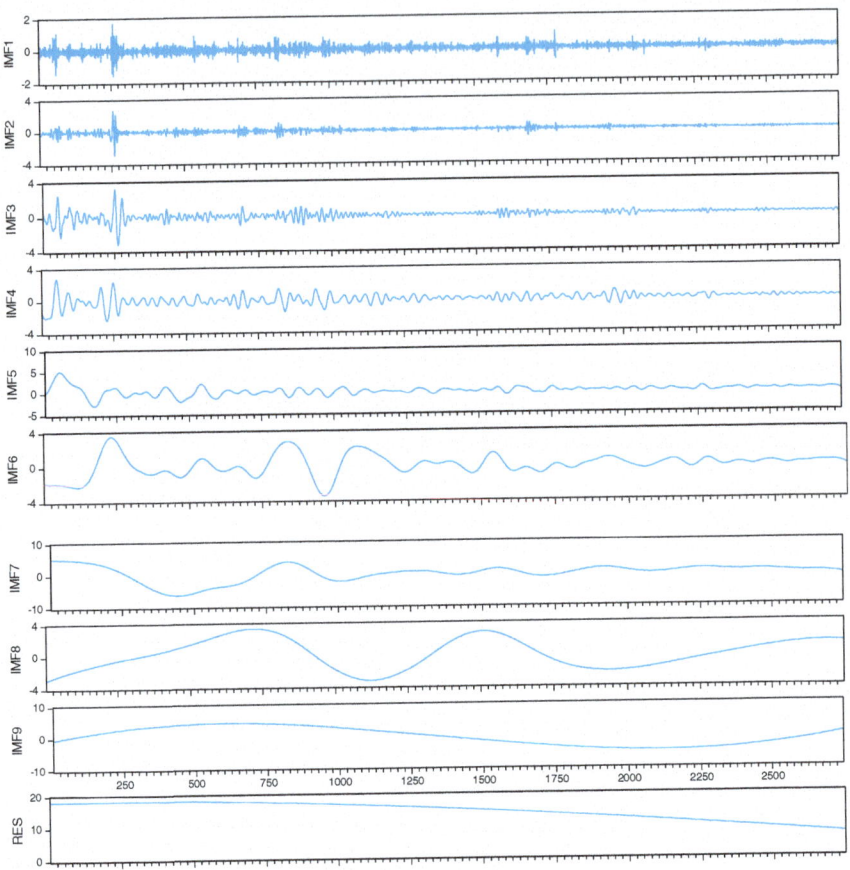

Fig. 4.2 IMFs and residue for the ECX daily data derived from EEMD

4.3.3 IMF Statistics

The following measures are taken to analyze IMFs: mean period of each IMF, correlation between each IMF (residue) and the original carbon price time series, the variance and variance percentage of each IMF. Related information is shown in Table 4.1.

The mean period here is defined as the value derived from dividing the total number of points by the number of peaks for each IMF since the frequency and the amplitude of an IMF may change with time continuously and the periods are also not constant. When an IMF is T long and has s peaks, the mean period of such an IMF is approximately: $t = T/s$. Two correlation coefficients, Pearson product moment correlation coefficient and Kendall rank correlation coefficient, are simultaneously employed in the analysis to measure the correlations between each IMF (residue) and the observed data from different points of view. Pearson

Table 4.1 Measures of IMFs and the residue for the ECX data derived from EEMD

	Mean period (day)	Pearson correlation	Kendall correlation	Variance	Variance as % of the observed	Variance as % of (ΣIMFs + residue)
Observed				49.641		
IMF_1	2.94	0.029	0.018	0.04	0.08	0.12
IMF_2	6.12	0.049**	0.038**	0.057	0.12	0.17
IMF_3	14.09	0.069**	0.034**	0.165	0.34	0.49
IMF_4	30.19	0.081**	0.032*	0.283	0.58	0.83
IMF_5	65.4	0.245**	0.171**	0.86	1.75	2.53
IMF_6	152.61	0.216**	0.075**	1.295	2.64	3.81
IMF_7	457.83	0.122**	0.007	5.485	11.19	16.15
IMF_8	915.67	0.335**	0.233**	3.289	6.71	9.68
IMF_9	1373.5	0.841**	0.667**	10.855	22.15	31.95
Residue	–	0.826**	0.585**	11.643	23.76	34.27
Sum					69.33	100

*Correlation is significant at the level of 0.05 (2-tailed); **Correlation is significant at the level of 0.01 (2-tailed)

correlation coefficient considers that, where the percentage that the value of IMF and carbon price series on one time point simultaneously exceed or are less than the mean value of series gets greater, the correlation coefficient gets higher, which is mainly used to measure the quantity correlation; Kendall coefficient considers that, where the percentage of changing in the same direction gets greater when comparing the value of IMF (residue) and carbon price series on one time point to those on the previous time point, the correlation coefficient gets higher, which is mainly used to measure the rank correlation. Two correlation coefficients measure the correlations between each IMF (residue) and the original carbon price series from different perspectives. Taking these two correlation coefficients into account at the same time, we can understand the correlation between each IMF (residue) and carbon price series at large. Since these IMFs (residue) are independent of each other, it is possible to sum up the variances and use the percentage of variance to explain the contribution of each IMF (residue) to the total volatility of the observed data. However, the variances of IMFs and the residue do not always add up to the observed variance, due to a combination of rounding errors, nonlinearity of the original time series, and introduction of variance by the treatment of the cubic spline end conditions (Peel et al. 2005; Zhang et al. 2008; Yu et al. 2015).

For the daily carbon price time series data decomposition, it is observed that the dominant mode of the observed data is not the residue but an IMF, i.e., IMF9. Both Pearson and Kendall coefficients, between the IMF9 and the observed data, reach a high level of more than 0.84 and 0.66 in Table 4.1, respectively, both with significance at the level of 0.01. At the same time, variance of the IMF9 accounts for nearly 22% of the total variability. Other important modes are the residue, IMF7 and IMF8, which have a mean period of more than one year. Interestingly, the two correlation coefficients differ very much for IMF7. This is because IMF7 varies slowly. An up (down) movement can last for a long time before the direction changes. Therefore, the fluctuation is sometimes reverse of the observed carbon price data, which is highly volatile. Although the residue and IMF9 also have the feature of a long cycle, their volatilities are not so highly, which makes their direction the same with that of the observed data at most of the data points; so the Kendall correlation for it is still high. Meanwhile, the sum of variances for these important components, IMF7, IMF8, IMF9, and the residue, contributes 63.81% of total variance. As Huang et al. (1998) mentioned, low-frequency modes are often treated as the indeterministic medium-term behavior, and the residue is often treated as the deterministic long-term behavior. On the other hand, the first IMFs not only exhibit very low correlation coefficients with the observed data but also account for about 0.1% of total variance. This means IMF1 does not have serious effect on carbon price, but reflects the short-term fluctuations of the two markets owing to the short-term unbalance of supply and demand (Zhang et al. 2008; Zhu 2012; Zhu et al. 2015). The higher the frequency of an IMF, the smaller its influence on the whole movement of carbon price.

4.4   Composition

In Sect. 4.3, decomposition results of ECX carbon price data sets and basic analysis of each IMF (residue) are provided. In this section, the IMFs are grouped into a high-frequency component and a low-frequency component, and the residue is recognized as a trend component based on the fine-to-coarse reconstruction algorithm as mentioned in Sect. 4.2.2. These three components have abundant economic meanings and reveal some new features of ECX carbon price.

According to the fine-to-coarse reconstruction algorithm, we calculated each t test value corresponding to the mean of $s_i(i = 1, 2, \ldots, 9)$ at the significance level of $\alpha = 0.05$, as shown in Table 4.2.

It can be found that the mean of the fine-to-coarse reconstruction departs significantly from zero at IMF4 at the significance level of $\alpha = 0.05$ for the first time, i.e., $i = 4$. For this reason, the partial reconstruction with IMF1, IMF2, and IMF3 represents the high-frequency component and the partial reconstruction with IMF4, IMF5, IMF6, IMF7, IMF8, and IMF9 represents the low-frequency component. The residue is treated separately as the trend component. Figure 4.3 shows the three components and Table 4.3 gives statistical measures, including Pearson and Kendall correlations between each component and the observed price, variance of each component and variance percentages.

Each component has some distinct characteristics. The residue, as mentioned before, is slowly varying around the long-term mean. Therefore, it is treated as the long-term trend during the evolution of EU ETS carbon price; each sharp up or down of the low-frequency component corresponds to a significant event, which should be a representative of the effect of these events; the high-frequency

Table 4.2 Mean of s_i and t value

	S1	S2	S3	S4	S5	S6	S7	S8	S9
Mean	−0.0001	−0.0008	−0.01	−0.0354	0.0706	0.0466	−0.0453	−0.0204	−0.343
t value	−0.0134	−0.1273	−0.9124	−2.0636	2.7994	1.4258	−0.7532	−0.3217	−3.8706
p value	0.9893	0.89868	0.3616	0.0391	0.0052	0.1541	0.4514	0.7477	0.0001

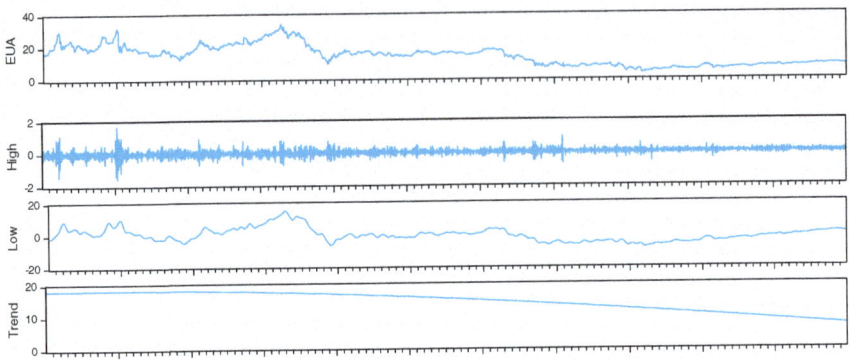

Fig. 4.3 Components of the ECX daily data from April 22, 2005 to January 12, 2016

Table 4.3 Correlations and variances

	Mean period (day)	Pearson correlation	Kendall correlation	Variance	Variance as % of the observed	Variance as % of (ΣIMFs + residue)
Observed				49.641		
High-frequency component	2.94	0.029	0.018**	0.04	0.08	0.12
Low-frequency component	39.24	0.908**	0.750**	21.172	43.21	64.44
Trend	–	0.826**	0.585**	11.64	23.76	35.44
Sum					67.05	100

**Correlation is significant at the level of 0.05 (2-tailed)

component, with the characteristics of small amplitudes, contains the effects of markets' short-term fluctuations and other irregular events.

For the movement trend of the original carbon price series, the most important component is low-frequency component. The Pearson and Kendall correlation coefficient between the low-frequency component and the original series reaches 0.908 and 0.75, respectively, both with significance at the level of 0.01; meanwhile, low-frequency component has higher relative variance contribution, amounting to 64.44%. Both high-frequency component and trend component have small effect on the whole movement of carbon price.

4.4.1   Trend

The trend is a crucial factor for the long-term movement of carbon price, but the long-term trend of the current carbon price holds a high correlation with the original carbon price and only 23.76% of volatility in the carbon price series is caused by the changes in trend. Notwithstanding the foregoing, the trend gives a reference to the long-term equilibrium price of carbon price, suggesting it is a deterministic force for carbon price evolution in the long run. The long-term equilibrium price is essentially the result from the outstanding performance of price discovery function after the international carbon market system is highly developed, and is shaped by the comprehensive game playing of all the stockholders associated with carbon market. The volatility of trend as a whole is consistent with the economic development of the world especially EU, which may imply that the long-term trend of ECX carbon price is determined by global economic development.

In fact, by comparing the trend with the observed carbon price, we can see that historically, although carbon price would fluctuate dramatically due to significant events, it would return to the trend after the influence of the event faded away. For example, the information disclosure of 2005 verified emissions in May 2006 made carbon price fall suddenly from 32.25 to 18.5 euros/ton, but the price rose slowly after that and finally returned to the trend price of 20 euros/ton or so in June 2006, the similar situation occurred during 2008–2009 and 2010–2012.

The trend appears to decline as a whole, which is related to the international events mentioned above. Moreover, the trend variance accounts for a small proportion of total variance, which is largely different from the changes in the whole trend of carbon price. Since the trend component changes in a clear direction, we can use a regression function with curve characteristics for the trend forecasting.

4.4.2   Effects of Significant Events

Due to the comparatively short period of development, EU ETS carbon market is positioned in a complicated external environment. Besides the effects of long-term trend and market mechanism, carbon price is also affected by significant events such as intergovernmental negotiations, national allocation plans (NAP), and global financial crisis. The effects of significant events are mainly embodied in the low-frequency component derived from IMF_7 to IMF_9. With regard to the mean periods of these IMFs, the shortest is nearly one year and the longest can be as long as 3 years, suggesting that it is hard for the market itself to eliminate these effects soon; the duration of the effect of a significant event maybe very long. In addition, the amplitudes at some carbon price data points could be more than 5 euros/ton or even higher, suggesting that the effects of some significant events on carbon price maybe very serious. Since the trend changes slowly and the market's normal fluctuation is small and happens at a high frequency, large fluctuations in medium term arise only from significant events.

Low-frequency component holds the highest correlation with the carbon price series, has a longer duration of a fluctuation, and is substantially identical with the general trend of carbon price volatility, so it has a higher Pearson correlation coefficient and Kendall correlation coefficient, arriving at 0.908 and 0.75, respectively. Compared with the high-frequency component, the mean of low-frequency component significantly deviates from zero and displays evident long-term memory. As a matter of fact, although the trend plays a crucial role in the trend of carbon price from a long-term perspective, it varies slowly and significant event is the most important cause to affect the movement of carbon price from a middle-term perspective. If we define the change rate as the difference between carbon prices for two consecutive days divided by the value of the earlier day, the change rate of the original carbon price is consistent with that of the low-frequency component. But this change rate for the low-frequency component often changes more slowly since it excludes the effects of activities occurring at a very high frequency. This is also the reason why the curve generated by the significant event looks like a smoothed carbon price time series.

By separating significant events included in the low-frequency component from the whole price, the effect of each significant event can be measured and then the result can be used as a reference for predicting the effect of the next significant event of the same type. For example, during the period of the information disclosure of 2005 verified emissions in May 2006, this component resulted in a price decrease of 14 euros/ton, which meant the maximum effect of this event was 14 euros/ton. This downward movement in price began at the beginning of May 2006 and the carbon price did not return to zero until the end of May 2006. Since no serious event occurred during this period, we can conclude that the effect of this event lasted for one month. However, the continued effects of this event did not disappear until February 2007. Under the global financial crisis beginning from September 2008, EU ETS carbon price fell to 10 euros/ton or below from the summit nearly 34 euros/ton in summer, with a decline of over 20 euros/ton, which roughly meant that the maximum effect of this event on carbon price was 20 euros/ton or above, and the effects of this event continue up to date.

Significant events under external environment generally take place infrequently, but once they break out, they may sharply shock the carbon price and push the carbon price to more than 10 euros/ton in a short time (information disclosure of verified emissions and European debt crisis) or even more than 20 euros/ton (global financial crisis); furthermore, they have a long duration. Taking the financial crisis in 2008 as an example, it lasts for more than 2 years and its effect on carbon price exists right now. Because the trend of carbon price varies slowly and the high-frequency data fluctuate in a small amplitude, carbon price varies largely only if being affected by external environment. Carbon market has limited ability of adjusting the price on its own, so that is why the volatility of low-frequency data is substantially consistent with the movement of carbon price. By analyzing the effects on carbon price, extent, and duration of historic significant events, we can seek some reference for the analysis of future similar events.

Low-frequency component presents well-disciplined fluctuations and makes great contribution to the carbon price volatility as a whole. In case of no extremely high requirement for the precision of forecasting, the direct extrapolation method may be applied to the forecasting, which is easy and practical. However, in case of any high requirement for precision, combination forecasting or other new approaches can be employed for carbon price forecasting.

4.4.3 Normal Market Disequilibrium

Besides significant events and the intrinsic trend, carbon prices are also influenced by many other factors, such as energy prices, industrial production, weather conditions, strikes, depletion of inventory, speculation fund operation, political activities, and some other stochastic volatility. Durations of these effects are often short. They are classified into high-frequency events and their effects are contained in the high-frequency component. Although we call this component as effects of normal market disequilibrium for short, it should be treated as a collection of events with short-term impact on carbon price. Since the ECX carbon price data used in our experiments are daily data, the words "short term" should be described, in general, as less than one month.

All IMFs from IMF_1 to IMF_3 have a shorter mean period and make little variance contribution, suggesting that the normal market fluctuations, such as supply–demand disequilibrium, have no serious impact on carbon price—it is generally not more than 3 euros/ton. Nonetheless, these events are becoming more and more frequent and have lately become the fundamental forces for pushing carbon prices up or down. Thus, normal market fluctuations can be neglected in long-term trend prediction, but they are important for short-term forecasting, besides that recent studies also find that these high-frequency IMFs contribute a lot to the interactions between carbon market and energy markets (Yu et al. 2015). Given the stochastic nature of high-frequency component, we can use time series-based models, such as ARIMA model which can well describe the stochastic process of carbon price time series, or artificial intelligence technologies, such as support vector machines (SVM) and artificial neural networks (ANN) which can well capture the nonlinear of carbon price time series, to carry out EU ETS carbon price forecasting so as to integrate the effects of various factors on such series.

The main finding of this section is that it is reasonable to assume that carbon price is composed of three components. The residue, also described as the "trend" in EEMD, represents the major trend of carbon price in the long run. The low-frequency component can be treated as the effects of significant events. It is the main reason for the dramatic carbon price variability in the medium term. However, the high-frequency component should be explained as normal market fluctuations or events which have only a short-term impact on carbon price. By virtue of this method, carbon price of 14 euros/ton or below in February 2007 can be decomposed into a trend price (about 18 euros/ton), effects of significant events (a number of significant events affecting the carbon market, such as information disclosure of verified emissions in May 2006 and inflow of speculative fund, with an influence size of about −4.2 euros/ton) and short-term market fluctuation (about 0.2 euros/ton); and carbon price of 10.8 euros/ton in February 2009 can be decomposed into a trend price (17 euros/ton), a significant event price (−6 euros/ton), and a normal fluctuation price (−0.2 euros/ton); similarly, carbon price of 11 euros/ton in October 2011 can be decomposed into a trend price (14 euros/ton), a significant event price (−3.1 euros/ton), and a normal fluctuation price (0.1 euros/ton).

4.5 Conclusion

The data of EU ETS ECX carbon future price are decomposed into several independent intrinsic modes with varying and different frequencies, bringing out some interesting features of carbon price volatility. The IMFs and the residue are summed up into only three components, based on fine-to-coarse reconstruction. Then the carbon price can be explained as the composite of a long-term trend, effect of significant event, and short-term fluctuations caused by normal supply–demand disequilibrium. Carbon price in the long run is basically determined by the trend, which changes continuously and hovers around the long-term mean. The sharp downs or ups in carbon prices are triggered by unpredictable and significant events, the impact of which may endure for several months and even longer. Otherwise, the small fluctuations in the short term are mainly driven by normal market activities or some small events which do not have serious effect on carbon markets.

By analyzing the composition of carbon price, we can consider some forecasting strategies: the first one is to predict each IMF based on its own characteristics, such as using a polynomial function to fit the residue, using a Fourier function to simulate low-frequency IMFs, and applying nonlinear forecasting technology to deal with high-frequency IMFs, and then integrate individual parts to obtain a final result. The second one might be grouping the IMFs into a nonlinear part and a linear part, forecasting each one individually, and then summing them up. In considering the concrete implications of each component, we suggest each part should be forecast based on both its concrete implications and data characteristics. The trend can be predicted by fitting the curve and the short-term fluctuations can be dealt with nonlinear forecasting techniques such as support vector regression and artificial neural network. However, it is difficult to predict and evaluate significant events. As everyone knows, a significant event itself is affected by many factors, such as political activities, weather, and other complicated factors. No one knows what will happen as well as when and where it will happen. Even though an irregular event may be expected to happen, evaluating its influence is still very difficult. Furthermore, even if the same event happens again, it may have different effects on carbon price at different times. There should be some new methods or an integrated forecasting framework to handle these issues. This chapter applies a novel approach, EEMD, to the analysis of carbon price and the resulting conclusion can be used as a guidance for carbon price forecasting.

References

Alberola E, Chevallier J, Cheze B (2008a) Price drivers and structural breaks in European carbon prices 2005–2007. Energy Policy 36:787–797

Alberola E, Chevallier J, Cheze B (2008b) The EU emissions trading scheme: the effects of industrial production and CO_2 emissions on European carbon prices. Int Econ 116:95–128

Alberola E, Chevallier J, Cheze B (2009) Emission's compliances and carbon prices under the EU ETS: a country specific analysis of industrial sectors. J Policy Model 31:446–462

Bataller MM, Pardo A, Valor E (2007) CO_2 prices, energy and weather. Energy J 28:73–92

Beat H (2010) Allowance price drivers in the first phase of the EUETS. J Environ Econ Manage 59:43–56

Bunn DW, Fezzi C (2008) A vector error correction model of the interactions among gas, electricity and carbon prices: an application to the cases of Germany and United Kingdom. In: Gulli F (ed) Markets for carbon and power pricing in Europe: theoretical issues and empirical analyses. Edward Elgar Publishing, pp 145–159

Chevallier J (2009) Carbon futures and macroeconomic risk factors: a view from the EU ETS. Energy Economics 31:614–625

Chevallier J, Ielpo F, Mercier L (2009) Risk aversion and institutional information disclosure on the European carbon market: a case-study of the 2006 compliance event. Energy Policy 31: 15–28

Christiansen A, Arvanitakis A, Tangen K et al (2005) Price determinants in the EU emissions trading scheme. Clim Policy 5:15–30

Cummings DAT, Irizarry RA, Huang NE et al (2004) Travelling waves in the occurrence of dengue haemorrhagic fever in Thailand. Nature 427:344–347

Feng ZH, Zou LL, Wei YM (2011) Carbon price volatility: evidence from EU ETS. Appl Energy 88:590–598

Huang NE, Shen Z, Long SR (1998) The empirical mode decomposition and the Hilbert spectrum for nonlinear and nonstationary time series analysis. Proc R Soc Lond A454:903–995

Huang NE, Shen Z, Long SR (1999) A new view of nonlinear water waves: the Hilbert spectrum. Annu Rev Fluid Mech 31:417–457

Huang NE, Wu ML, Qu WD et al (2003) Applications of Hilbert Huang transform to non-stationary financial time series analysis. Appl Stoch Models Bus Ind 19:245–268

Kanen JLM (2006) Carbon trading and pricing. Environmental Finance Publications, London

Keppler JH, Bataller MM (2010) Causalities between CO_2, electricity, and other energy variables during phase I and phase II of the EU ETS. Energy Policy 38:3329–3341

Koenig P (2011) Assessing rollover criteria for EUAs and CERs. Int J Econ Finan Issues 4: 669–676

Montagnoli A, de Vries FPD (2009) Carbon trading thickness and market efficiency. Stirling Econ Discuss Pap 32(6):1331–1336

Paolella MS, Taschini L (2006) An econometric analysis of emission trading allowances. Research paper series, Swiss Finance Institute, Zurich

Peel MC, Amirthanathan GE, Pegram GGS et al (2005) Issues with the application of empirical mode decomposition. In: Zerger A, Argent RM (eds) Modsim 2005 international congress on modelling and simulation, pp 1681–1687

Reilly JM, Paltsev S (2005) An analysis of the European emission trading scheme. In: MIT Joint Program on the Science and Policy of Global Chang, Report No. 127

Rilling G, Flandrin P, Goncalves P (2003) On empirical mode decomposition and its algorithms. In: IEEE EURASIP Workshop on Nonlinear Signal and Image Processing Grado(1)

Rittler D (2012) Price discovery and volatility spillovers in the European Union Emissions trading scheme: a high-frequency analysis. J Bank Finance 36(3):774–785

Seifert J, Marliese UH, Michael W (2008) Dynamic behavior of CO_2 spot prices. J Environ Econ Manage 56:180–194

Springer U (2003) The market for tradable GHG permits under the Kyoto Protocol: a survey of model studies. Energy Econ 25:527–551

Uhrig-Homburg M, Wagner M (2009) Futures prices dynamics of CO_2 emission allowances: an empirical analysis of the trial period. J Deriv 17:73–88

Wu Z, Huang NE (2009) Ensemble empirical mode decomposition: a noise-assisted data analysis method. Adv Adapt Data Anal 1(1):1–41

Yu L, Wang SY, Lai KK et al (2010) A multiscale neural network learning paradigm for financial crisis forecasting. Neurocomputing 73:716–772

Yu L, Li J, Tang L, Wang S (2015) Linear and nonlinear Granger causality investigation between carbon market and crude oil market: a multi-scale approach. Energy Econ 51:300–311

Zhang D (2008) Oil shock and economic growth in Japan: a nonlinear approach. Energy Econ 30 (5):2374–2390

Zhang YJ, Wei YM (2010) An overview of current research on EU ETS: evidence from its operating mechanism and economic effect. Appl Energy 87:1804–1814

Zhang X, Lai KK, Wang SY (2008) A new approach for crude oil price analysis based on empirical mode decomposition. Energy Econ 30:905–918

Zhu B (2012) A novel multiscale ensemble carbon price prediction model integrating empirical mode decomposition, genetic algorithm and artificial neural network. Energies 5(2):355–370

Zhu B, Wang P, Chevallier J, Wei Y (2015) Carbon price analysis using empirical mode decomposition. Comput Econ 45(2):195–206

Chapter 5
Modeling the Dynamics of European Carbon Futures Prices: A Zipf Analysis

Abstract This chapter proposes a Zipf-type analysis in order to understand the trading behavior on the carbon market. Several categories are devised to characterize greedy versus non-greedy speculators. We map the trading horizons of buy-and-hold (bearish) long-term investors versus short-term speculators taking advantage of short-term (bullish) extreme price movements.

5.1 Introduction

The carbon market is not only an important tool for human beings to address climate change, but also an important choice for investors to spread their investment risks. Carbon pricing has become one of the key issues of global carbon market developments. At present, whether by market value or trading volume, the EU ETS carbon futures market is the world's largest, having become a guide to the global carbon market. Therefore, this work focuses on the EU ETS carbon futures price behavior analysis.

In recent years, carbon price behavior has attracted attention from scholars. Mansanet-Bataller and Valor (2007), Alberola et al. (2008), and Creti et al. (2012) used multiple linear regression models with dummy variables to investigate changes in carbon futures price behavior. Paolella and Taschini (2008) and Chevallier (2009) used the GARCH model to examine the volatility of carbon futures price from 2005 to 2007. Benz and Truck (2008) used the MS–AR–GARCH model to explore the volatility of carbon futures price returns between 2005 and 2007. Chevallier (2010, 2011), and Conrad et al. (2012), respectively, used the FAVAR, HAR–RV, and FIAPGARCH models to model the volatility of the carbon futures price. Feng et al. (2011) used a nonlinear method to analyze the volatility of carbon futures price returns. Zhu (2012) used empirical mode decomposition (EMD) combined with an artificial neural network (ANN) modeling to predict the EU ETS carbon futures price. Zhu and Wei (2013) introduced least squares support

Special thanks to Shujiao Ma, Lili Yuan and Ying-Ming Wei for collaborating research on Chap. 5.

vector machines (LSSVM) into their EU ETS carbon futures price forecasts, which over performed ARIMA and ANN models. These studies provide us with important references to understand the volatility of carbon futures price; however, they rarely make full use of the information contained in the carbon futures price fluctuations (up or down) to explore the volatility of carbon futures price at different investment timescales and different speculators' expectations of returns. Thus, they cannot provide comprehensive information to support decision-making.

In this article, we apply the Zipf analysis for the EU ETS carbon futures price to explore the dynamic behavior of carbon futures price at different investment timescales, and following different speculators' expectations of returns. By doing so, we obtain abundant information about carbon futures price fluctuations. The Zipf analysis, initially used in the chapter of natural language (Zipf 1949, 1968), has been applied to examine the behavior of financial markets in recent years, for more details see Refs. (Axtell 2001; Jiao et al. 2006; He et al. 2006; Vandewalle and Ausloos 1999; Alvarez-Ramirez 2003; Xiao and Wang 2012). It has also been applied recently in the field of applied statistics by Niu and Wang (2013). However, it has not been yet applied to the analysis of carbon markets to date. We fill this gap in the literature.

The Zipf analysis maps the carbon price fluctuations, upward and downward, using 1 and −1, respectively, to generate a new string sequence containing fundamental information of price fluctuations to investigate the dynamic behavior of the original price sequence. When the carbon price dynamic behavior follows a random walk, a probability of 1 should be the same as that of −1. When the probability of 1 is not same as that of −1, this indicates that carbon price dynamic behavior has the characteristics of a trend and long-term memory. It appears beneficial when trying to understand the price formation mechanism of carbon futures, so as to better avoid carbon price risk, and to better inform future transactions.

5.2 Methodology

5.2.1 Zipf Analysis

Zipf analysis, originally introduced in the context of natural languages, has been applied to various types of data in physical and social sciences. To study the fluctuation of carbon price changes, in this section, we apply the Zipf analysis method to investigate the symbolic dynamics of real data from the EU ETS carbon futures market.

Let $p(t) = \{p(t_1), p(t_2), \ldots, p(t_n)\}$ denote the original carbon price time series, and $r_i(\tau)$ the returns of day i under timescale τ. The definition of τ-returns of carbon price is given by formula (5.1):

$$r_i(\tau) = [p(t_i + \tau) - p(t_i)]/p(t_i), i = 1, 2, \ldots, n - \tau, \tag{5.1}$$

from which we get the τ-return series $r_i(\tau) = \{r_1(\tau), r_2(\tau), \ldots, r_{n-\tau}(\tau)\}$. Next, we map the τ-return series into a new three-alphabet symbolic sequence $f_i(\tau, \varepsilon)$ as formula (5.2):

$$f_i(\tau, \varepsilon) = \begin{cases} -1, & \text{if } ri < -\varepsilon \\ 0, & \text{if } -\varepsilon \leq ri \leq \varepsilon. \\ 1, & \text{if } ri < -\varepsilon \end{cases} \qquad (5.2)$$

We obtain a new sequence which contains the most basic information of carbon price upward, stable, and downward moves $f(\tau, \varepsilon) = \{f_1(\tau, \varepsilon), f_2(\tau, \varepsilon), \ldots, f_{n-\tau}(\tau, \varepsilon)\}$, with τ a given timescale. Especially, when $\tau = 1, 5, 20, 60, 120$, or 250, these are called characteristic timescales, which approximately stand for one transaction day (TD), one transaction week, one transaction quarter, one transaction half year, and one transaction year, respectively, in terms of business time (units with weekends and holidays eliminated). 1, 0, and -1 denote "price-up", "price-stable", and "price-down", respectively. ε is a price variation threshold, which can be interpreted as the expected returns for investors.

To further explore the information of carbon futures price behaviors, we introduce absolute frequencies and relative frequencies to examine carbon price behaviors of $f(\tau, \varepsilon)$ for different values of parameters ε and τ. We let $n_+(\tau, \varepsilon)$, $n_0(\tau, \varepsilon)$, and $n_-(\tau, \varepsilon)$ denote the frequencies of occurrences for price-up, price-stable, and price-down, respectively, and $n_-(\tau, \varepsilon) + n_0(\tau, \varepsilon) + n_+(\tau, \varepsilon) = n - \tau$. Then the corresponding absolute frequencies of sequence $f_i(\tau, \varepsilon)$ are given by formula (5.3):

$$p_-(\tau, \varepsilon) = n_-(\tau, \varepsilon)/n - \tau$$
$$p_0(\tau, \varepsilon) = n_0(\tau, \varepsilon)/n - \tau \qquad (5.3)$$
$$p_+(\tau, \varepsilon) = n_+(\tau, \varepsilon)/n - \tau.$$

The corresponding relative frequencies are written as formula (5.4):

$$\Phi_-(\tau, \varepsilon) = n_-(\tau, \varepsilon)/n_\pm(\tau, \varepsilon)$$
$$\Phi_+(\tau, \varepsilon) = n_+(\tau, \varepsilon)/n_\pm(\tau, \varepsilon), \qquad (5.4)$$

where $n_\pm(\tau, \varepsilon) = n_-(\tau, \varepsilon) + n_+(\tau, \varepsilon)$. In the definition of the relative frequencies, we neglect occurrences of price-stability, and use $\Phi_+(\tau, \varepsilon)$ and $\Phi_-(\tau, \varepsilon)$ to measure the total occurrences of price rise and price fall, respectively. In the following discussion, for the actual carbon price data, we consider the statistical properties of absolute frequencies and relative frequencies for different values of the two parameters ε and τ.

5.2.2 Economic Significance of the Parameters ε and τ

In the EU ETS market, there are different types of speculators, such as suppliers, energy consumers, governments, banks, and so on. Among them, their respective influences, derived from speculators' behavior and psychology, are difficult to quantitatively analyze and forecast. Different investment timescales lead to different expectations. The parameter τ is designed to reflect the average transaction time interval of speculators. The higher the τ, the longer the corresponding transaction time interval for the speculators. The parameter $\varepsilon(\varepsilon \geq 0)$ is designed as the speculators' psychological threshold to expected returns. The maximum earnings expectation is $+\varepsilon$, and the maximum risk value is $-\varepsilon$. To simplify, we assume that the absolute values of $\pm\varepsilon$ are equal. Usually, speculators believe that the current carbon futures price fluctuation depends on their subjective psychological threshold ε. Due to the presence of risks and costs caused by market transaction costs and uncertainties, speculators believe that when returns are higher than $+\varepsilon$, European carbon futures price shows a substantial rise. After reaching their prospective earnings, they will sell their own European Union Allowances (EUAs). When returns are between $\pm\varepsilon$, speculators believe that the European carbon futures price has not undergone a substantial change. Due to prevalent transaction costs, they will not participate in market trading and choose to continue to hold negatively. When returns are inferior to $-\varepsilon$, speculators believe that the European carbon futures price is about to undergo a substantial decrease. Hence, some of them opt for the buy-and-hold strategy. The parameter ε reflects speculators' psychological endurance concerning European carbon futures price fluctuations, as well as their optimism about future market trends.

From the perspective of speculators, short-term speculators need to watch for each peak and trough in the sequence of European carbon futures price variations, which constitutes their profit pattern. Their frequency of attention to the carbon market is much higher than long-term investors. Long-term investors' concern is the long-term trend of European carbon futures price fluctuations, thus their frequency of attention to the market is lower, and their value of τ is therefore higher. According to whether European carbon futures price achieves speculators' desired returns, they correspondingly make their decisions. Usually, when the European carbon futures price does not match speculators' expected returns or risk levels, it produces trading desire. Not all speculators' decisions are rational, but when making so-called irrational decisions, speculators are changing their expected returns.

5.3 Empirical Analyses

5.3.1 Data

In this article, we use the EUA carbon futures price data from the ECX, the most liquid carbon futures market under the EU ETS, from April 22, 2005 to January 12,

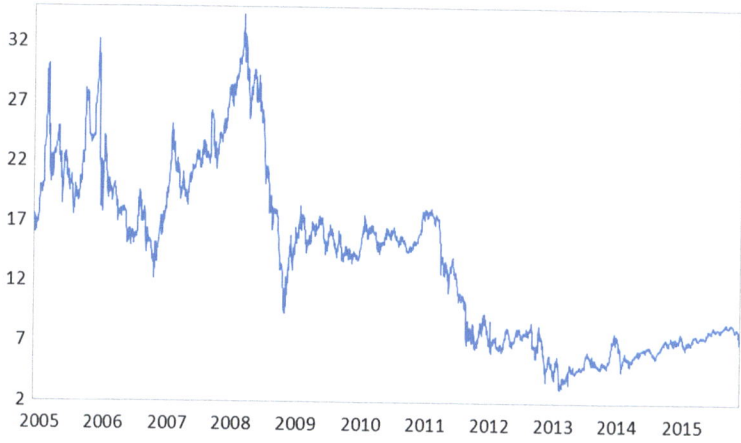

Fig. 5.1 European carbon futures price during April 22, 2015–January 12, 2016

2016, totalling of 2747 daily observations, as shown in Fig. 5.1. According to Eqs. (5.1)–(5.4), we preprocess the raw data and obtain the returns sequence of carbon prices, the upward or downward sequences of carbon prices under different values of ε and τ, as well as the absolute and relative frequencies of each sequence, respectively.

5.3.2 The Influences of Investment Timescale and Investor Psychology on the Expected Returns

According to Eq. (5.2), we map the returns sequences data of carbon futures price into string sequences, and then conduct a preliminary analysis of the new sequences. We conduct the deviation (namely 1 frequency minus −1 frequency) analysis under different investment timescales and speculators' anticipations: the results are shown in Fig. 5.2.

When only considering investment timescale variations, without considering speculators' expected returns, i.e., $\varepsilon = 0$, results unfold as follows. First, because the deviation in each timescale is nonzero, the investment timescale has important effects on carbon futures price fluctuations.

Second, although in the extreme case, $\varepsilon = 0$ and $\tau = 1$, we observe a deviation of 0.00364 between the probabilities of carbon futures price-upward and price-downward moves. This implies that the EU ETS carbon futures price fluctuations are asymmetric.

Third, there is a deviation of less than zero within each characteristic timescale, and the deviation increases with further oscillations.

Fourth, speculators' expectations for carbon futures price-upward moves are greater than carbon futures price-downward moves. With the extension of the

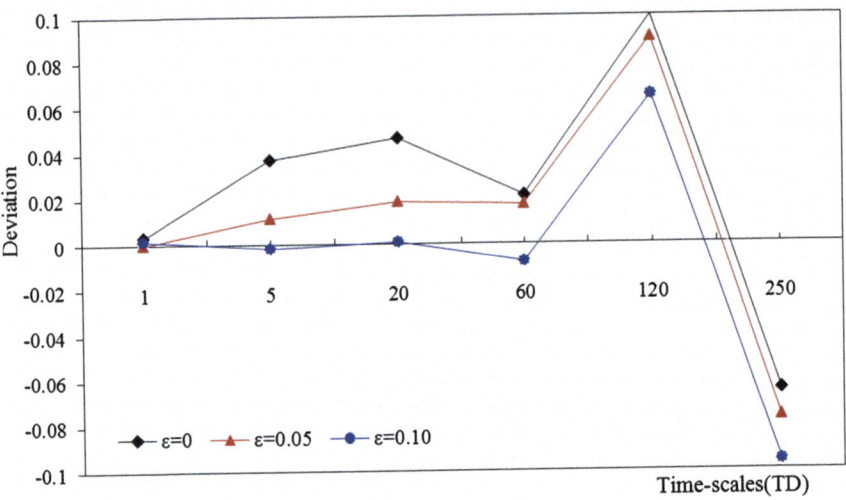

Fig. 5.2 The deviations of probabilities of carbon price-up and price-down under characteristic timescales

investment period, speculators generally believe that carbon futures price would decrease in the long term.

When considering the speculators' expected returns, i.e., $\varepsilon > 0$, we can draw the following conclusions. First, speculators' expectations of returns have dramatic influences on carbon futures price fluctuations.

Second, differing expectations of returns ($\varepsilon = 0.05$ and $\tau = 1$) cause deviations under different timescales.

Third, with increased timescale τ, deviations decrease below those corresponding to the absence of consideration of psychological expectations' characteristic timescales. This result implies that speculators' expectations may distort carbon price fluctuations for a certain period, further promoting price decreases.

This preliminary analysis reveals that investment timescales and speculators' expectations of returns influence the carbon futures price behavior. How can we determine the investment timescales and measure the effect of different timescales on the carbon futures price behavior? How can we measure the effect of speculators' expectations of returns on the carbon futures price behavior? To answer these questions, further work is needed.

5.3.3 Division of Speculators Based on Parameters

With the increase of speculators' expectations of returns, how do the absolute and relative frequencies change? We compute the changes of absolute and relative frequencies under different $\varepsilon (0 \leq \varepsilon \leq 1)$ by increments of 0.05: the corresponding results are shown in Table 5.1, and Figs. 5.3 and 5.4.

Table 5.1 Various frequencies' critical points $\varepsilon_c(\tau)$

$\varepsilon_c(\tau)$	$p_-(\tau,\varepsilon)$	$p_0(\tau,\varepsilon)$	$p_+(\tau,\varepsilon)$	$\Phi_-(\tau,\varepsilon)$	$\Phi_+(\tau,\varepsilon)$
$\tau = 1$	0.30	0.35	0.35	0.30	0.30
$\tau = 5$	0.45	0.45	0.45	0.45	0.45
$\tau = 20$	0.5	0.6	0.6	0.5	0.5
$\tau = 60$	0.55	0.95	0.95	0.55	0.55
$\tau = 120$	0.7	0.85	0.85	0.65	0.65
$\tau = 250$	0.65	0.95	0.95	0.7	0.7

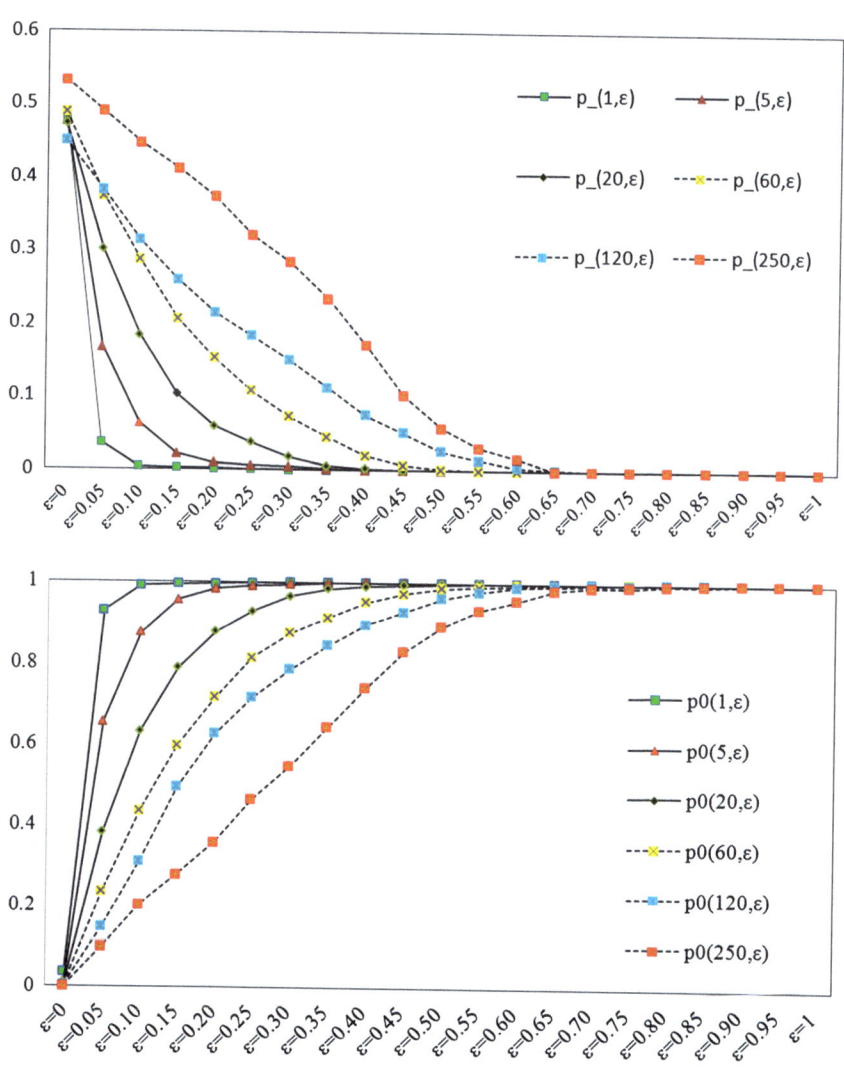

Fig. 5.3 The evolution of ε and saturation of absolute frequencies

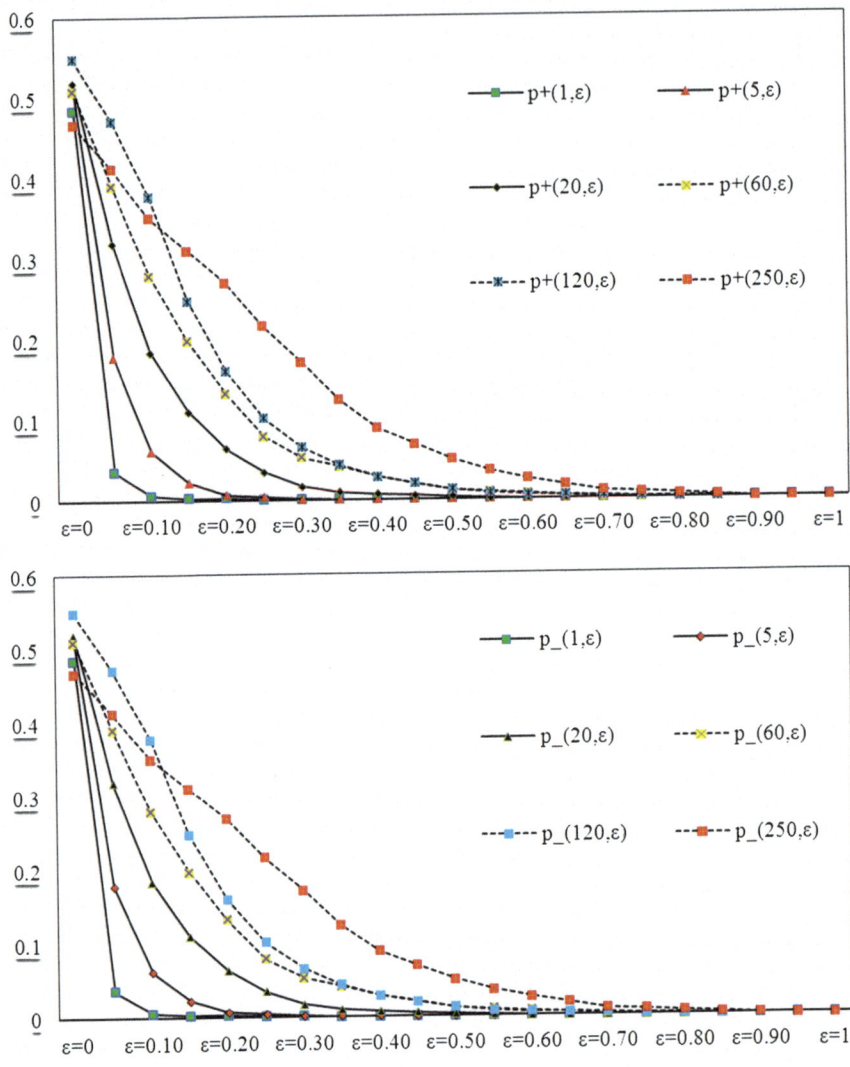

Fig. 5.3 (continued)

Figure 5.2 shows that speculators with different expectations of returns may have different perceptions of carbon futures price fluctuations, so it is necessary to classify speculators and explore the characteristics of different categories thereof. Figures 5.3 and 5.4 show that, with the evolution of ε, the absolute and relative frequencies tended to be saturated respectively, and the concrete critical points, namely $\varepsilon_c(\tau)$, are presented in Table 5.1. Each frequency has reached saturation after critical points.

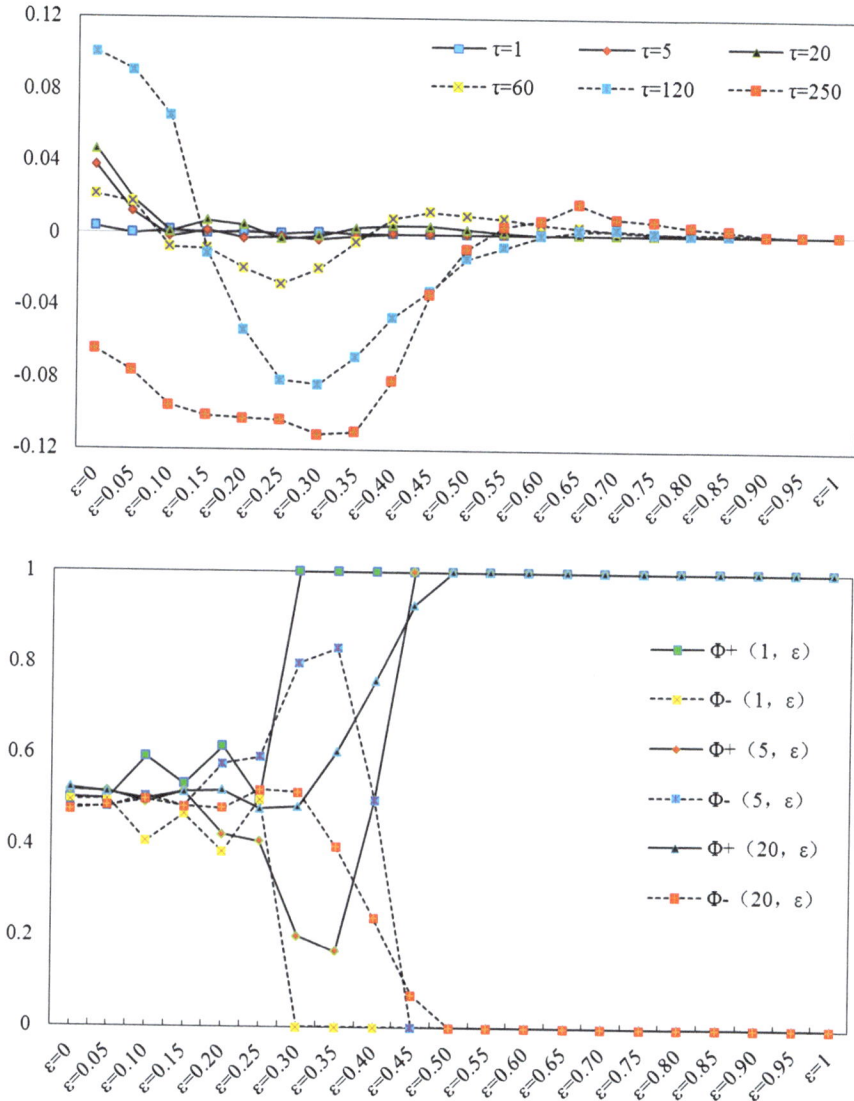

Fig. 5.3 (continued)

Although speculators' expectations of returns can the distort carbon futures price, this distortion is not indefinite. As with the growth of expectations of returns, there is a saturation point. After saturation, frequencies of increase and decrease remain stable. This means that once speculators' expectations of returns reach a certain level, they will no longer be able to distort carbon futures price behavior.

Under different expectations of returns, we obtain the following results concerning the carbon price behavior. First, the higher the speculators' expectations of

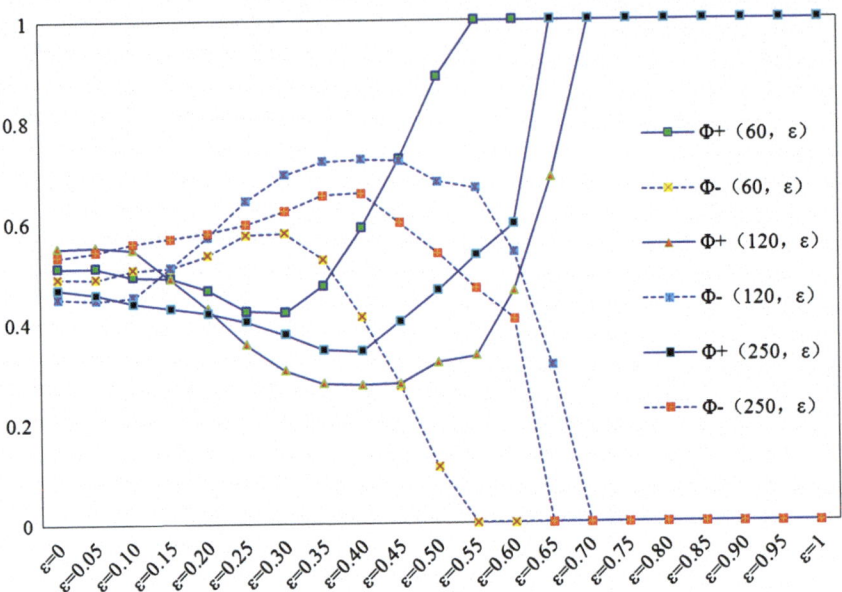

Fig. 5.4 The saturation of relative frequencies and division of speculators

returns, the more extreme the effect on the judgment of carbon price behavior. For instance, when $\tau = 20$ and $\varepsilon < 0.25$, the carbon price behavior is almost random. When $\varepsilon \geq 0.25$, the probability of the carbon price going up or down is almost equal to zero.

Second, the deviations of carbon price-upward and price-downward movements in different expectations of returns are different at the same investment timescales. Under the same timescale, with the improvement of expectations of returns, the deviations become smaller, until reaching saturation.

Third, there is an asymmetry to carbon price fluctuations. Before reaching the saturation point, the deviation of carbon price-upward and price-downward moves is nonzero, but the probability of increasing is slightly higher than that of decreases. For example, with a lower expectation of returns ($\varepsilon = 0.05$) and a middle timescale of investment (120 TD), the probability of increases is higher than that of decreases by 9.03%.

According to Fig. 5.4, speculators can be roughly divided into two categories: greedy ($\varepsilon \geq 0.25$) and non-greedy ($\varepsilon < 0.25$) speculators. For speculators who have smaller timescale investments ($\tau = 1, 5$, or 20), when $\varepsilon < 0.25$ (non-greedy), the relative frequencies are approximately 0.5, and the probability of the carbon price going up is similar to that of carbon price going down, i.e., the carbon price behavior is close to a random walk. However, when $\varepsilon \geq 0.25$ (greedy), the relative frequencies quickly converge to the extreme values of 0 or 1. For speculators who have longer timescales for their investment ($\tau = 60, 120$, or 250), the relative frequencies also show changes from slow to high growth before and after $\varepsilon = 0.25$.

Table 5.2 Actual historical carbon price information in characteristic timescales

Frequency	Timescales (TD)					
	$\tau = 1$	$\tau = 5$	$\tau = 20$	$\tau = 60$	$\tau = 120$	$\tau = 250$
$p_-(\tau, 0)$	0.48034	0.47610	0.47322	0.48809	0.44842	0.53165
$p_0(\tau, 0)$	0.03569	0.01058	0.00697	0.00261	0.00284	0.00120
$p_+(\tau, 0)$	0.48398	0.51332	0.51981	0.50931	0.54874	0.46715
$\Phi_-(\tau, 0)$	0.49811	0.48120	0.47654	0.48936	0.44976	0.53230
$\Phi_+(\tau, 0)$	0.50189	0.51881	0.52346	0.51064	0.55021	0.46771

5.3.4 Absolute Frequencies of Carbon Price Fluctuations

According to Eq. (5.3), we obtain the absolute frequencies of carbon price, namely $p_-(\tau, \varepsilon)$, $p_0(\tau, \varepsilon)$, and $p_+(\tau, \varepsilon)$. Figure 5.3 shows how the absolute frequencies evolve for each particular timescale of investment τ, when speculators' expectations of returns are equal to $\varepsilon = 0.05, 0.1, 0.2, 0.3, 0.4$, and 0.5. To quantify speculators' cognition of the carbon price behavior, and distortions from the current carbon price behavior, we take speculators' expectation of returns $\varepsilon = 0$ as a reference, namely taking $p_-(\tau, 0)$, $p_0(\tau, 0)$, and $p_+(\tau, 0)$ as the references, and $p_-(\tau, \varepsilon) - p_-(\tau, 0)$ as the distortion for speculators' cognition of the historical carbon price information. The corresponding results are shown in Table 5.2. Figure 5.5 shows that there are some distortions in speculators' cognitions of the historical carbon price behavior.

Greedy and non-greedy speculators have similar cognitions of historical carbon price information. In contrast, Fig. 5.5 shows that non-greedy speculators have the complete opposite cognition of carbon price-downward and price-stable conditions. Figure 5.5a shows that non-greedy speculators' cognition of the carbon price-downward move is rising. Figure 5.5b shows that non-greedy speculators' cognition of the carbon price-stable is falling. In the meantime, Fig. 5.5c shows that speculators' cognitions of carbon price-downward frequencies exhibit inflection points. The corresponding data are shown in Table 5.3.

Table 5.4 shows that the distortion caused by speculators' expectation of returns to $p_0(\tau, 0)$ is basically positive. In contrast, Tables 5.5 and 5.6 show that the distortions caused by speculators' expectation of returns to $p_-(\tau, 0)$ and $p_+(\tau, 0)$ are both negative. Figure 5.5 and Tables 5.4, 5.5 and 5.6 show that with increased investment timescales, the distortion caused by speculators' expectation of returns generally tends to decrease.

5.3.5 Relative Frequencies of Carbon Price Fluctuations

According to Eq. (5.4), we obtain the relative frequencies of carbon price variations, as shown in Figs. 5.6 and 5.7. Due to $\Phi_+(\tau, \varepsilon) + \Phi_-(\tau, \varepsilon) = 1$, $\Phi_-(\tau, \varepsilon)$ and $\Phi_-(\tau, \varepsilon)$ being symmetrical, we only discuss $\Phi_+(\tau, \varepsilon)$. Similar to Sect. 3.4, we

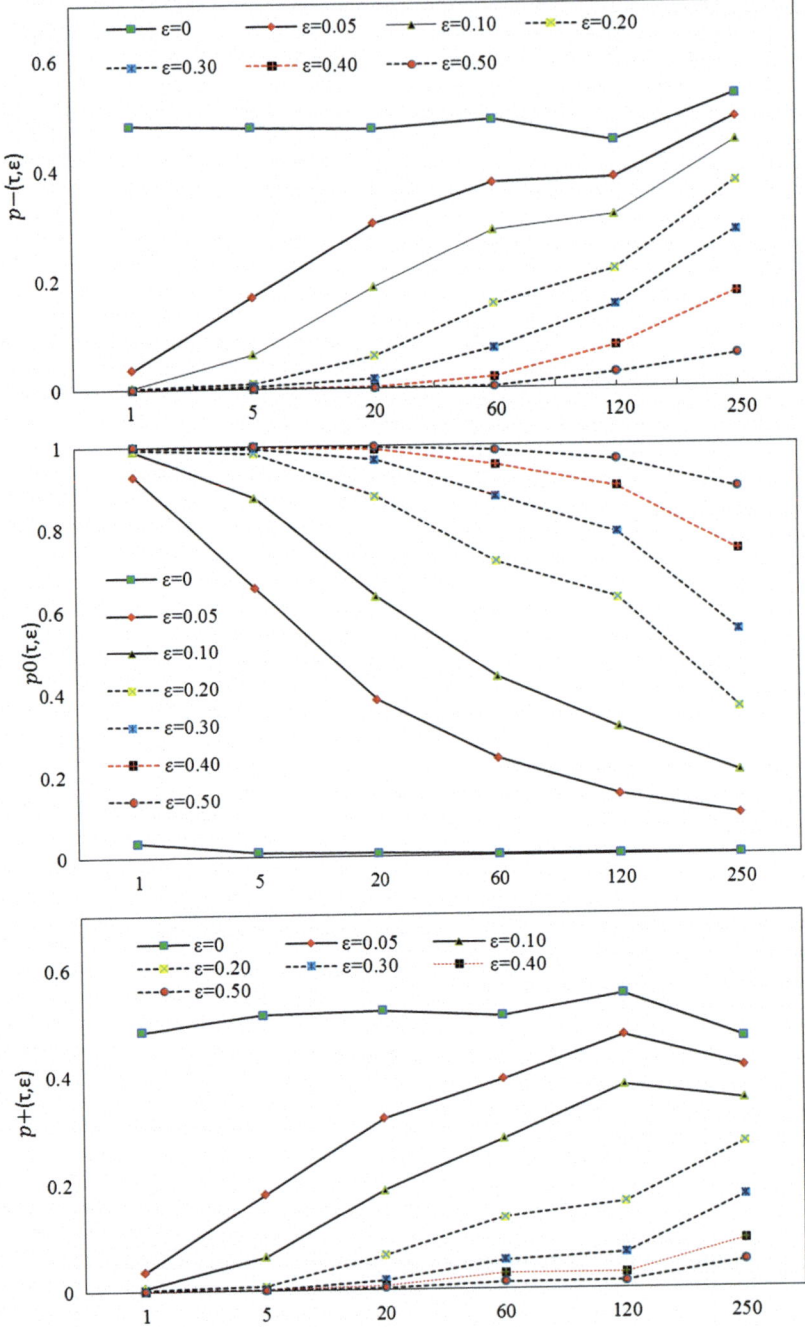

Fig. 5.5 Speculators' cognitions of historical carbon price information (absolute frequencies)

Table 5.3 The inflection points $\varepsilon_c(\tau)$ of $p_+(\tau, \varepsilon)$ in non-greedy expectations

	$\varepsilon = 0$	$\varepsilon = 0.05$	$\varepsilon = 0.1$	$\varepsilon = 0.15$	$\varepsilon = 0.2$
$\varepsilon_c(\tau)$	120	120	120	250	250

Table 5.4 The information distortion of speculators' expectations relative to $p_0(\tau, 0)$ in characteristic timescales

Absolute frequency	Timescales (TD)					
	$\tau = 1$	$\tau = 5$	$\tau = 20$	$\tau = 60$	$\tau = 120$	$\tau = 250$
$\varepsilon = 0.05$	0.892207	0.643561	0.373441	0.232688	0.144536	0.096955
$\varepsilon = 0.1$	0.954479	0.864283	0.624725	0.431869	0.30676	0.201122
$\varepsilon = 0.2$	0.959578	0.973003	0.870139	0.712584	0.622068	0.355769
$\varepsilon = 0.3$	0.963583	0.983947	0.958914	0.872673	0.782388	0.544071
$\varepsilon = 0.4$	−0.03569	−0.01022	−0.00477	0.017126	0.072183	0.169471
$\varepsilon = 0.5$	−0.03569	−0.01058	−0.00697	−0.00112	0.023058	0.05649

Table 5.5 The information distortion of speculators' expectations relative to $p_-(\tau, 0)$ in characteristic timescales

Absolute frequency	Timescales (TD)					
	$\tau = 1$	$\tau = 5$	$\tau = 20$	$\tau = 60$	$\tau = 120$	$\tau = 250$
$\varepsilon = 0.05$	−0.44428	−0.30901	−0.17278	−0.1143	−0.06723	−0.04247
$\varepsilon = 0.1$	−0.47633	−0.41262	−0.28944	−0.20141	−0.13578	−0.08494
$\varepsilon = 0.2$	−0.47851	−0.46662	−0.41416	−0.33582	−0.23403	−0.15865
$\varepsilon = 0.3$	−0.48034	−0.47173	−0.45561	−0.41586	−0.29914	−0.2484
$\varepsilon = 0.4$	−0.48034	−0.47574	−0.47102	−0.46835	−0.3734	−0.36098
$\varepsilon = 0.5$	−0.48034	−0.4761	−0.47322	−0.4866	−0.42253	−0.47396

Table 5.6 The information distortion of speculators' expectations relative to $p_+(\tau, 0)$ in characteristic timescales

Absolute frequency	Timescales (TD)					
	$\tau = 1$	$\tau = 5$	$\tau = 20$	$\tau = 60$	$\tau = 120$	$\tau = 250$
$\varepsilon = 0.05$	−0.44792	−0.33455	−0.20066	−0.11839	−0.0773	−0.05449
$\varepsilon = 0.1$	−0.47815	−0.45166	−0.33529	−0.23045	−0.17098	−0.11619
$\varepsilon = 0.2$	−0.48106	−0.50638	−0.45598	−0.37677	−0.38804	−0.19712
$\varepsilon = 0.3$	−0.48325	−0.51222	−0.5033	−0.45681	−0.48324	−0.29567
$\varepsilon = 0.4$	−0.48398	−0.51295	−0.51284	−0.48101	−0.52018	−0.37821
$\varepsilon = 0.5$	−0.48398	−0.51332	−0.51724	−0.49739	−0.53656	−0.41747

investigate the distortion of speculators' expectation of returns to the historical carbon price information in terms of relative frequencies. The results are reported in Table 5.7. In contrast with absolute frequencies, the distortion caused by speculators' expectation of returns is mostly negative. Only a fraction of them is positive,

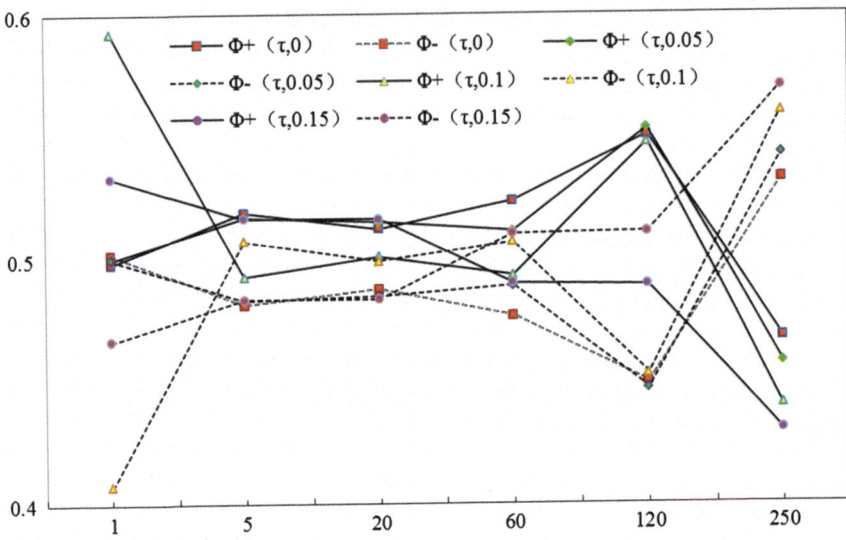

Fig. 5.6 The historical price information cognitions of non-greedy speculators (relative frequencies)

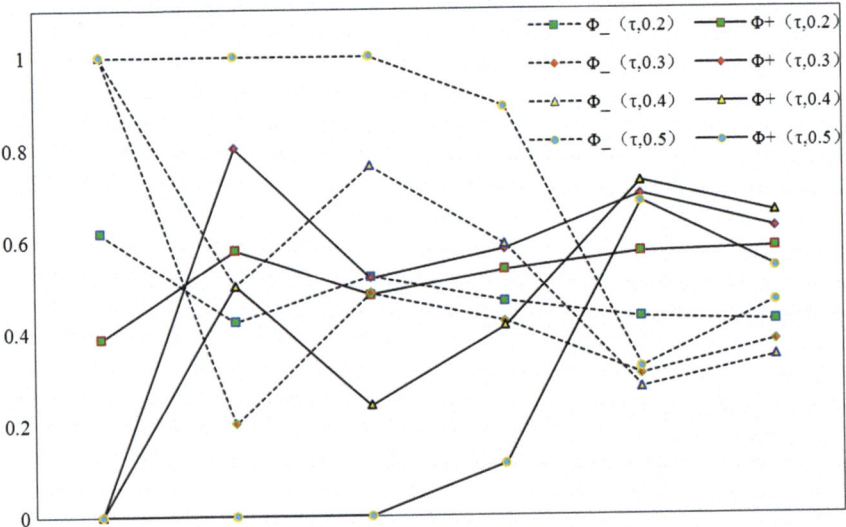

Fig. 5.7 The historical price information cognitions of greedy speculators (relative frequencies)

and negative distortions are more likely to occur over longer ($\tau \geq 120$) than smaller timescales ($\tau \leq 5$).

Figures 5.6 and 5.7 show that non-greedy and greedy speculators have entirely different cognitions of the historical carbon price information in terms of relative

Table 5.7 The information distortion of speculators' expectations relative to $\Phi_+(\tau,0)$ in characteristic timescales

$\Phi_+(\tau,\varepsilon)$	Timescales (TD)					
	$\tau = 1$	$\tau = 5$	$\tau = 20$	$\tau = 60$	$\tau = 120$	$\tau = 250$
$\varepsilon = 0.05$	−0.00189	−0.00193	−0.00836	0.000559	0.002715	−0.01014
$\varepsilon = 0.1$	0.090704	−0.02609	−0.02246	−0.01755	−0.00305	−0.02773
$\varepsilon = 0.2$	0.113496	−0.09658	−0.00405	−0.04528	−0.12178	−0.04777
$\varepsilon = 0.3$	0.498112	−0.31881	−0.03959	−0.08974	−0.24525	−0.09062
$\varepsilon = 0.4$	0.498112	−0.01881	0.236542	0.078509	−0.27447	−0.12512
$\varepsilon = 0.5$	0.498112	0.481195	0.476542	0.378251	−0.23021	−0.00502

frequencies. The former cognition patterns are generally similar to actual historical information. The latter cognitions patterns differ completely from the historical information.

Under lower expected returns, carbon price fluctuations are relatively well understood. When $\varepsilon \leq 0.25$, the probabilities of the carbon price going up and down are accurate, following two stages. During the first stage, carbon prices fluctuate more slowly, and relative probabilities are found between 0.4 and 0.6. Thus, carbon price fluctuations correspond to the market's spontaneous behavior as a result of market mechanism functions. During the second stage, we observe impacts from heterogeneous events. The global financial crisis in 2008 and the European debt crisis in 2011 led to a reduction in manufacturing output, a corresponding reduction in EUA demand, and the carbon price fell sharply. Although the influence of these heterogeneous events is in long-term, their probability of occurrence was small.

Under higher expected returns, the carbon price fluctuations are not accurate. When $\varepsilon \geq 0.25$, the relative probability becomes relatively scattered and quickly converges to 0 or 1. The probabilities of the carbon price going up and down are so unstable that they lead to large deviations. Investment returns become unstable. As a consequence, speculators' expected returns cannot be too high.

5.4 Results: Analysis and Discussion

Due to speculators' decisions being based on their cognitions of the historical information, we discuss the influences of speculators' trading time intervals and expected returns on the European carbon futures price formation mechanism following our Zipf analysis.

First, at a specific timescale, the absolute and relative frequencies converge to their saturation points. Although speculators can distort the carbon price behavior, this distortion will not induce unlimited growth. With the growth of expectations, whether considering absolute or relative frequencies, there is a saturation point. The frequency is no longer growing beyond, implying that once speculators' expectations reach a certain higher level, they will no longer be able to distort the carbon price behavior. Owing to carbon price fluctuations, they cannot meet speculators' unreasonably higher expectations. Unless speculators modify their expectations, it is disadvantageous to trade. In addition, short-term speculators' ($\tau \leq 20$) trend judgments on carbon price fluctuation tend to extremes. When $\tau = 1$, non-greedy speculators tend to believe that the carbon price is close to a random walk, and it becomes difficult to judge any trend therein, whereas greedy speculators argue that the possibility of carbon price going up is 100%.

Second, in terms of absolute frequencies, different expected returns have distorting effects on carbon price behavior. The higher the speculators' expectations, the larger the distortion of real historical information. Further, the distortions to $p_0(\tau, 0)$ are mostly positive, while for other frequencies distortions are negative. With increased timescale, the distortions for $p_+(\tau, 0)$ and $p_0(\tau, 0)$ tend to increase. The cognitions of non-greedy speculators for historical carbon price information exhibit similarities and internal consistencies with current carbon price trends. Although speculators' expectations distort the carbon price behavior, they cannot affect the long-term movement of carbon prices. The differences in market cognitions arising from different expectations of speculators are mainly amplitudes and occasions of carbon price fluctuations, rather than carbon price fluctuations themselves. Speculators' cognitions for long-term carbon price-downward moves tend to be more consistent. They believe that the probability of carbon price-stability reduces with increased investment timescales. However, speculators' cognitions for carbon price-upward moves have obvious inflection points. Speculators' cognitions have fundamental changes before and after inflection points. Before inflection points, speculators with different expected returns believe that the carbon price exhibits a likelihood of rising. The lower the expected returns, the greater the confidence in a carbon price short-term increase. After inflection points, they usually believe that the possibility of the carbon price going up is smaller, and they cannot make more profits. For speculators expecting lower returns, there is a high probability that the carbon price would rise in the short term. However, in the long term, this probability tends to decline, i.e., with the extension of the investment timescale, speculators tend to believe that carbon price is bearish.

Because two types of speculators have different cognitions of the carbon market, we discuss non-greedy and greedy speculators separately with regards their relative frequencies. For non-greedy speculators, the curve of $\Phi_+(\tau, \varepsilon)$ lays mostly above 0.5, and the curve of $\Phi_-(\tau, \varepsilon)$ lay mostly under 0.5. With increased τ, $\Phi_+(\tau, \varepsilon)$ decreases, which means that the carbon price behavior is biased and exhibits a negative trend. The long-term bearish probability is bigger, which corroborates the aforementioned deviation analysis of absolute frequencies. Speculators' expectations have distorting effects on the carbon price behavior. When $\varepsilon = 0.1$, the probability of carbon price-downward move is less than 0.5 within about a trade week ($\tau \leq 5$), and the probability of carbon price-up is less than 0.5. $\Phi_+(\tau, \varepsilon) \geq 0.5$ and $\Phi_-(\tau, \varepsilon)$ has obvious inflection points. Short- and medium-term investments ($1 \leq \tau \leq 120$) show changing probability of carbon price nearing 0.5, and stabilizing between 0.4 and 0.6. At this moment, carbon price fluctuations are relatively close to a random walk, and the probability of carbon price going up is the same as that of the carbon price going down. Hence, long-buying and short-selling becomes a potentially profitable trading strategy. Permanent investment ($\tau > 120$) exhibits an unstable carbon price behavior, as the probability of carbon decreases becomes important. When the probability of carbon price-upward moves greatly decreases, speculators tend to believe that carbon price is bearish in the long term. Although there remains the possibility for short-term bullishness, speculators tend to believe that the carbon price will be bearish in the long term. When $\varepsilon = 0.1$ and $\tau \leq 5$, there were $\Phi_+(\tau, \varepsilon) \geq 0.5$ and $\Phi_+(\tau, \varepsilon) \leq 0.5$, which differed from the price perceptions of speculators who believed that $\varepsilon = 0.15$ or 0.2.

For greedy speculators, in the short-term, although speculators have volatile perceptions of carbon price fluctuations, and the greedier the speculator, the greater the volatility, the curve of $\Phi_-(\tau, \varepsilon)$ lays mostly below 0.5, and the curve of $\Phi_+(\tau, \varepsilon)$ lays mostly above 0.5. With increased τ, $\Phi_-(\tau, \varepsilon)$ increases, which means that the carbon price behavior is biased and exhibits a negative trend, i.e., the long-term bearish probability is higher, which is also consistent with non-greedy speculators. Greedy speculators and non-greedy speculators have entirely different perception patterns of short-term carbon price fluctuations. Speculators with $\varepsilon = 0.3$ believe that there are great deviations between 0 and 1 during $1 \leq \tau \leq 5$, which shows that greedy speculators' risk propensity is so strong that they are irrational. Speculations based on this acknowledged facet of the carbon market are likely to aggravate carbon price volatility, market risk, and uncertainty. However, aggravated carbon price volatility is likely to be expected by higher expectations speculators, if they want to achieve their expectations. Overall, the bearish probability of these speculators on the carbon price is still higher than their bullish probability.

5.5 Concluding Remarks

Based on our Zipf analysis, we can draw the following conclusions. First, the carbon price behavior is asymmetric. Whatever speculators' expectations and speculative timescales, the long-term bearish probability is, on the whole, greater than the long-term bullish probability.

Second, timescales of investment and speculators' expectations of returns have dual effects on the carbon price behavior. The longer the timescales of investment, the higher the bearish probability. The lower the expected returns, the smaller the distortion of the carbon price behavior. Speculators who have average transaction time intervals less than 20 tend to two extremes in their acknowledgement of carbon price fluctuations: non-greedy speculators believe that the carbon price fluctuation is close to a random walk, whereas greedy speculators believe that the carbon price is either 100% bullish or 100% bearish. Speculators who have average transaction time intervals over 60 are relatively less extreme. They believe that the carbon price will remain bearish in the long-term, and with the growth of their expected returns, that the bearish probability slowly increases.

Third, the differences in carbon market cognitions from non-greedy speculators with different expectations mainly lay in the amplitudes and occasions of carbon price fluctuations, rather than carbon price fluctuations themselves. Compared with non-greedy speculators, greedy speculators have entirely different perception patterns of the carbon market, being very unstable and tending to go to extremes. When speculating based on their acknowledged perception of the carbon market, they will greatly distort the carbon price and aggravate its volatility, therefore increasing market uncertainty.

Fourth, once speculators' expected returns have reached critical points, they will no longer be able to distort carbon price behavior, which means that unless speculators modified their higher expectations, they risked being, pardon the colloquialism, wiped out of the carbon market.

Fifth, for non-greedy-type speculators ($\varepsilon < 0.25$), in the short term ($1 \leq \tau < 60$), the probabilities of carbon price up and down are pretty much the same, namely carbon price behavior is relatively close to random walk. Due to the existence of transaction costs, they will not participate in the market trading and choose continue to hold negatively. In the long term ($\tau \geq 60$), the probability of carbon price-downward movements is declining slowly, and the probability of carbon price-upward movements is increasing slowly. As the carbon price tends to fall in the long term, investors can choose a buy-and-hold trading strategy. For greedy-type speculators ($\varepsilon \geq 0.25$), and their expected returns are in the range of critical point, in the short term ($1 \leq \tau < 60$), the ups probability of carbon price quickly converges to 1, and the downs probability of carbon price quickly converges to 0. When the carbon price substantially rises, the investor can sell his European Union Allowances (EUAs). In the long term ($\tau > 60$), the ups probability of carbon price quickly converges to 0, and the downs probability of carbon price quickly converges to 1. When the carbon price records a dramatic fall, the price signal tends to decline in the long-term, and therefore the investor can opt for the buy-and-hold trading strategy.

References

Alberola E, Chevallier J, Chèze B (2008) Price drivers and structural breaks in European carbon prices 2005–2007. Energy Policy 36(2):787–797

Alvarez-Ramirez J, Soriano A, Cisneros M et al (2003) Symmetry/anti-symmetry phase transitions in crude oil markets. Phys A 322:583–596

Axtell RL (2001) Zipf distribution of U.S. firm sizes. Science 293:1818–1820

Benz E, Truck S (2008) Modeling the price dynamics of CO_2 emission allowances. Energy Econ 31:4–15

Chevallier J (2009) Carbon futures and macroeconomic risk factors: a view from the EU ETS. Energy Economics 31:614–625

Chevallier J (2010) Volatility forecasting of carbon prices using factor models. Econ Bull 30:1642–1660

Chevallier J (2011) Nonparametric modeling of carbon prices. Energy Econ 33:1267–1282

Conrad C, Rittler D, Rotfub W (2012) Modeling and explaining the dynamics of European Union Allowance prices at the high-frequency. Energy Economics 34(1):316–326

Creti A, Jouvet PA, Mignon V (2012) Carbon price drivers: Phase I versus Phase II equilibrium? Energy Econ 34(1):327–334

Feng ZH, Zou LL, Wei YM (2011) Carbon price volatility: evidence from EU ETS. Appl Energy 88:590–598

He LY, Fan Y, Wei YM (2006) Empirical study on Zipf analysis of Brent crude oil price behaviours. Complex Syst Complex Sci 3:67–78

Jiao JL, Fan YL, Wei YM et al (2006) Study on gasoline prices behaviour based on Zipf technique. Syst Eng Theor Pract 10:44–49

Mansanet-Bataller M, Valor E (2007) CO_2 prices energy and weather. Energy J 28(3):73–92

Niu H, Wang J (2013) Power-law scaling behavior analysis of financial time series model by voter interacting dynamic system. J Appl Stat 40(40):313–360

Paolella MS, Taschini L (2008) An econometric analysis of emission allowance prices. J Bank Finance 32:2022–2032

Vandewalle N, Ausloos M (1999) The N-Zipf analysis of financial data series and biased data series. Phys A 268:240–249

Xiao D, Wang J (2012) Modeling stock price dynamics by continuum percolation system and relevant complex systems analysis. Phys A 391:4827–4838

Zhu BZ (2012) A novel multiscale ensemble carbon price prediction model integrating empirical mode decomposition, genetic algorithm and artificial neural network. Energies 5:355–370

Zhu BZ, Wei YM (2013) Carbon price prediction with a hybrid ARIMA and least squares support vector machines methodology. Omega 41:517–524

Zipf GK (1949) Human behavior and the principle of least effort. Addison-Wesley Press, Cambridge

Zipf GK (1968) The psycho-biology of language: an introduction to dynamic psychology. Addison-Wesley Press, Cambridge

Chapter 6
Carbon Price Forecasting with a Hybrid ARIMA and Least Squares Support Vector Machines Methodology

Abstract This chapter advances a hybrid forecasting model for the carbon market. The technology is based on Least Squares Support Vector Machines augmented by particle swarm optimization (PSO). This innovation reaches superior forecasting results in a horse-race containing several combinations of ARIMA time series models.

6.1 Introduction

As the first multinational atmospheric greenhouse gas cap-and-trade system, the EU ETS is the largest carbon market in the world up to date, which has proven to be not only an important tool for mankind to address climate changes, but also a major choice for investors to decentralize their investment risks (Wei et al. 2010; Zhang and Wei YM 2010). Therefore, the need for more accurate forecasts of carbon prices is driven by the desire to reduce risk and uncertainty.

Recently, although carbon price analysis has become one of the key issues concerned by many academic researchers and business practitioners (Wei et al. 2010; Seifert et al. 2008; Bataller et al. 2007; Beat 2010; Keppler and Bataller 2010), only a few articles regarding carbon price forecasting can be found. In fact, carbon price changes over time, which can be treated as a time series process. Therefore, carbon price forecasting is a kind of time series forecasting. During the past few decades, various approaches have been developed for time series forecasting, among which the autoregressive integrated moving average (ARIMA) model has been found to be one of the most popular time series forecasting models due to its statistical properties, as well as the well-known Box-Jenkins methodology in the modeling process. However, the ARIMA model is only a class of linear model and thus it can only capture linear patterns of a time series. Therefore, the ARIMA model cannot effectively capture any nonlinear patterns hidden in a time series.

In order to overcome the limitations of the linear models and account for the nonlinear patterns existing in real cases, numerous nonlinear models have been

Special thanks to Lili Yuan and Ying-Ming Wei for supporting writing of Chap. 6.

proposed, among which the artificial neural network (ANN) has shown excellent nonlinear modeling capability. Although a large number of successful applications have shown that ANN can be successfully adopted in many forecasting fields (Shafie-khah et al. 2011; Firat and Gungor 2009; Zhang and Wu 2009; Ioannou et al. 2009), ANN suffers from some weaknesses, such as locally optimal solutions and over-fitting, which will still make the forecasting precision unsatisfactory. In 1995, a novel ANN, support vector machines (SVM), was developed by Vapnik (1995). Established on the structural risk minimization (SRM) principle by minimizing an upper bound of the generalization error, SVM can result in resistance to the over-fitting problem (Vapnik et al. 1997). However, SVM formulates the training process through quadratic programming, which can take much more time. In 1999, Suykens and his colleagues proposed a novel SVM known as least squares support vector machines (LSSVM) (Suykenns and Vandewalle 1999), which is able to solve linear problems quicker with a straightforward approach. Until now, LSSVM has been successfully used in pattern recognition and nonlinear regression estimation problems. At the same time, to obtain the optimal LSSVM model, it is important to choose a kernel function and determine the kernel parameters. Therefore, we introduce particle swarm optimization (PSO) (Kennedy and Eberhart 1995) to optimize the parameters of LSSVM in this chapter. Although both LSSVM and ARIMA models have achieved success in their own linear or nonlinear domains, neither is suitable for all circumstances. The approximation of ARIMA models to complex, nonlinear problems, as well as LSSVM to model linear problems, may be inappropriate, let alone in problems that simultaneously contain both linear and nonlinear patterns. Since it is difficult to completely know the characteristics of a real situation, a hybrid methodology that has both linear and nonlinear modeling capabilities can be a good candidate for practical use (Zhang 2003), which has been demonstrated by numerous studies (Zhang 2003; Khashei and Bijari 2011; Jain and Kumar 2007; Omer 2009; Koutroumanidis et al. 2009; Kong and Wu 2008; Pai and Lin 2005; Chen and Wang 2007; Cagdas et al. 2009; Tseng et al. 2002). However, existing hybrid methodologies often combine the traditional ANN (Zhang 2003; Khashei and Bijari 2011; Jain and Kumar 2007; Omer 2009; Koutroumanidis et al. 2009; Kong and Wu 2008) or SVM (Pai and Lin 2005; Chen and Wang 2007; Cagdas et al. 2009; Tseng et al. 2002) and ARIMA models, no LSSVM and ARIMA hybrid model has been found for carbon price forecasting, and this chapter thus aims to fill this gap.

The contributions made by this chapter may be twofold. First, we establish a novel LSSVM and ARIMA hybrid forecasting methodology to forecast carbon prices. In our proposed methodology, carbon prices are decomposed into two components: a linear component and a nonlinear component. An ARIMA model and a LSSVM model is used to capture the linear and nonlinear components of carbon prices respectively, and their forecasting values are integrated into the final forecasting results. Furthermore, PSO is used to find the optimal LSSVM parameter settings to forecast the carbon prices in the future. Second, we evaluate the forecasting performance of the single ARIMA and LSSVM models, the combined ARIMA and LSSVM model (COMBINED model), and the hybrid ARIMA and

LSSVM model, for forecasting the carbon prices under the EU ETS. The empirical results obtained demonstrate that the proposed hybrid model can outperform the single ARIMA, LSSVM models, and the COMBINED model in terms of statistical measures and trading performances.

6.2 Methodology

6.2.1 ARIMA Model

In the ARIMA model, carbon price is a linear function of past values and error terms. An ARIMA (p, d, q) model of degree of AR (p), difference (d), and MA (q) can be mathematically expressed as formula (6.1):

$$x_t = u_t + \phi_1 x_{t-1} + \phi_2 x_{t-2} + \cdots + \phi_p x_{t-p} - \theta_1 \varepsilon_{t-1} - \theta_2 \varepsilon_{t-2} - \cdots - \theta_q \varepsilon_{t-q} \quad (6.1)$$

where x_t is carbon price obtained by differencing d times, ε_t (hypothetical white noise) is assumed to be independently and identically distributed with a mean of zero and a constant variance of σ_ε^2, p, and q refer to the number of autoregressive and moving average terms in the ARIMA model, and $\phi_i(i = 1, 2, \ldots, p)$, $\theta_i(i = 1, 2, \ldots, p)$ are model parameters to be estimated.

Fitting an ARIMA model to carbon price involves the following four-step iterative processes: determine the structure of the ARIMA model, estimate the parameter values of the ARIMA model, perform ARIMA model tests on the residuals, and predict future carbon prices.

The major advantage of the ARIMA model is that it can capture the linear patterns of carbon prices well and is relatively easy to use. However, the ARIMA, on its own, is not adequate for carbon price forecasting because real carbon prices are often nonlinear and irregular. Therefore, we introduce the LSSVM model to capture nonlinear patterns existing in the carbon price.

6.2.2 Least Squares Support Vector Machines for Regression

In contrast to other forecasting approaches, SVM, firstly proposed by Vapnik (1995) in 1995 and based on the SRM principle, has been successfully applied to classification and regression. However, SVM training is a time-consuming process when analyzing huge data. LSSVM is a modification of the standard SVM developed by Suykens et al. Suykenns and Vandewalle (1999) in 1999 to overcome these shortcomings, which results in a set of linear equations instead of a quadratic programming problem. Consider a given training set $\{x_i, y_i\}, i = 1, 2, \ldots, l$ with input data, x_i, and output data, y_i. LSSVM defines the regression function as formula (6.2):

$$\min J(w, e) = \frac{1}{2} w^T w + \frac{1}{2} \gamma \sum_{i=1}^{l} e_i^2 \tag{6.2}$$

subject to formula (6.3):

$$y_i = w^T \varphi(x_i) + b + e_i, i = 1, 2, \ldots, l \tag{6.3}$$

where ω is the weight vector, γ is the penalty parameter, e_i is the approximation error, $\varphi(\cdot)$ is the nonlinear mapping function and b is the bias term. The corresponding Lagrange function (6.4) can be obtained

$$L(w, e, a, b) = J(w, e) - \sum_{i=1}^{l} a_i \{ w^T \varphi(x_i) + b + e_i - y_i \} \tag{6.4}$$

where a_i is the Lagrange multiplier. Using the Karush–Kuhn–Tucker (KKT) conditions, the solutions can be obtained by partially differentiating with respect to w, b, e_i, and a_i as formula (6.5):

$$\begin{cases} \frac{\partial L}{\partial w} = 0 \rightarrow & w = \sum_{i=1}^{l} a_i \varphi(x_i) \\ \frac{\partial L}{\partial b} = 0 \rightarrow & \sum_{i=1}^{l} a_i = 0 \\ \frac{\partial L}{\partial e_i} = 0 \rightarrow & a_i = \gamma e_i \\ \frac{\partial L}{\partial a_i} = 0 \rightarrow & w^T \varphi(x_i) + b + e_i - y_i = 0. \end{cases} \tag{6.5}$$

By eliminating w and e_i, the equations can be changed into formula (6.6):

$$\begin{bmatrix} b \\ \alpha \end{bmatrix} = \begin{bmatrix} 0 & I_v^T \\ I_v & \Omega + \gamma^{-1} I \end{bmatrix}^{-1} \begin{bmatrix} 0 \\ y \end{bmatrix} \tag{6.6}$$

where $y = [y_1, y_2, \ldots, y_l]^T$, $I_v = [1, 1, \ldots, 1]^T$, $\alpha = [\alpha_1, \alpha_2, \ldots, \alpha_l]^T$, and the Mercer condition (Vapnik 1995) has been applied to matrix Ω with $\Omega_{km} = \varphi(x_k)^T \varphi(x_m), k, m = 1, 2, \ldots, l$. Therefore, the LSSVM for regression can be obtained as formula (6.7):

$$y(x) = \sum_{i=1}^{l} \alpha_i K(x, x_i) + b \tag{6.7}$$

where $K(x, x_i)$ is the kernel function.

A major advantage of the LSSVM model is that it can capture the nonlinear patterns hidden in the carbon price. However, using LSSVM model alone to model the carbon prices may produce mixed results (Taskaya and Casey 2005; Yu et al. 2005). Therefore, we can conclude that the relationship between the ARIMA and LSSVM models is complementary, and it is necessary to combine the two for effective carbon price forecasting. And we define the predicted value of COMBINED model is equal to the sum of the predicted values of the LSSVM method and the ARIMA model divided by two.

6.2.3 The Hybrid Models

In reality, carbon price is rarely purely linear or nonlinear, but often contains both linear and nonlinear patterns, due to its inherently high complexity. This makes carbon price forecasting very difficult. Neither ARIMA nor LSSVM can sufficiently model and predict the carbon price since the linear model cannot deal with nonlinear patterns while the nonlinear model alone cannot handle both linear and nonlinear patterns equally well (Taskaya and Casey 2005; Yu et al. 2005). Hybridizing the two models can yield a robust method and result in good forecasting results. Therefore, motivated by the previous studies (Zhang 2003; Khashei and Bijari 2011; Jain and Kumar 2007; Omer 2009; Koutroumanidis et al. 2009; Kong and Wu 2008; Pai and Lin 2005; Chen and Wang 2007; Cagdas et al. 2009; Tseng et al. 2002), we propose a novel hybrid methodology which integrates the ARIMA and LSSVM models that can further improve forecasting accuracy.

The carbon price $\{Y_t, t = 1, 2, \ldots, n\}$ is considered as the sum of a linear component and a nonlinear component, i.e., $Y_t = L_t + N_t$, where L_t and N_t are the respective linear and nonlinear components to be estimated. First, an ARIMA model is used to fit the linear component of the carbon price, and generate a series of forecasts defined as $\{\hat{L}_t\}$. Next, the residuals containing only the nonlinear patterns in the carbon price, denoted as $\{\varepsilon_t\}$, can be generated from comparing the actual value Y_t of the carbon price with forecast value $\hat{L}_t$ of the linear component. That is, $\varepsilon_t = Y_t - \hat{L}_t$, where $\hat{L}_t$ is the forecast value of the ARIMA model at time t. Through modeling the residuals with a LSSVM, nonlinear patterns can be discovered. Finally, we build three various hybrid models with the following inputs:

Model 1: ARIMALSSVM1

$$\hat{N}_t = f(\varepsilon_{t-1}, \varepsilon_{t-2}, \ldots, \varepsilon_{t-p}) \tag{6.8}$$

where $\hat{N}_t$ is the forecast value of nonlinear component of the carbon price at time t, f is a nonlinear regression function determined by the LSSVM model and p is determined by the ARIMA model. Therefore, the final forecasting results will be calculated as (6.9):

$$\hat{Y}_t = \hat{L}_t + \hat{N}_t \tag{6.9}$$

Model 2: ARIMALSSVM2

$$\hat{Y}_t = f(Y_{t-1}, Y_{t-2}, \ldots, Y_{t-p}, \varepsilon_t) \tag{6.10}$$

where f is a nonlinear regression function determined by the LSSVM model and p is determined by the ARIMA model.

Model 3: ARIMALSSVM3

$$\hat{Y}_t = f(Y_{t-1}, Y_{t-2}, \ldots, Y_{t-p}, \hat{L}_t) \tag{6.11}$$

where f is a nonlinear regression function determined by the LSSVM model and p is determined by the ARIMA model.

6.3   The Optimal LSSVM Model by Particle Swarm Optimization

In LSSVM, choosing inappropriate parameters can result in over-fitting or under-fitting (Chen and Wang 2007; Cagdas et al. 2009). To build a LSSVM model efficiently, model parameters must be accurately determined (Kavaklioglu 2011). These parameters include:

(1) Kernel function: The kernel function is used to construct a nonlinear decision hyper-surface on the LSSVM input space. In general, using the RBF kernel function, $K(x, y) = \exp(\frac{-\|x-y\|^2}{2\sigma^2})$, can yield a good prediction (Smola 1998). Therefore, we adopt the RBF kernel function as the kernel function of the LSSVM model.
(2) The regularization parameter, γ, balances the complexity and approximation accuracy of the LSSVM model.
(3) The bandwidth of the kernel function, σ, describes the variance of the RBF kernel function.

PSO is the one of the modern heuristic algorithms developed by Kennedy and Eberhart in 1995 (Kennedy and Eberhart 1995). PSO uses several particles to constitute a swarm, and each particle represents a potential solution. Each particle flies around within the search space to seek the global optimal solution. In this chapter, we propose a new model known as PSO-LSSVM to optimize both the LSSVM parameters in order to improve the prediction accuracy. Figure 6.1 illustrates the processes of the proposed PSO-LSSVM model, which is described in detail as follows:

Step 1: *Data preparation*. The carbon price data are separated into the training and testing sets known as Tr and Te, respectively.

Step 2: *Particle initialization and PSO parameter setting*. The two LSSVM parameters, γ and σ are directly coded with real values to randomly generate a swarm of initial particles. The PSO parameters are set, including the number of particles m particle dimension (D), maximal iterations (t_{max}), position limitation (p_{max}), velocity limitation (v_{max}), inertia weight limitation [w_{min}, w_{max}], acceleration coefficients c_1 and c_2, γ limitation [γ_{min}, γ_{max}], and σ limitation [σ_{min}, σ_{max}]. Therefore, the position and velocity of the i-th particle can be treated as $x_i = (x_{i1}, x_{i2}, \ldots, x_{iD})$ and $v_i = (v_{i1}, v_{i2}, \ldots, v_{iD})$, respectively. The i-th particle has its own best position $p_{best} = (p_{i1}, p_{i2}, \ldots, p_{iD})$ where the best fitness encountered by the particle so far, and the global best position is denoted by $g_{best} = (p_{g1}, p_{g2}, \ldots, p_{gD})$ for all particles in the current generation. The iterative variable is set to $t = 0$ and the training process can begin.

Step 3: *Fitness definition*. Because the training error can be measured by the root mean squared error (RMSE) built with a given set of parameters $\{\gamma, \sigma\}$, we can define the fitness function as formulas (6.12) and (6.13):

$$MinF_{fitness} = RMSE(\gamma, \sigma) \tag{6.12}$$

$$RMSE(\gamma, \sigma) = \sqrt{\frac{1}{n}\sum_{i=1}^{n}[y_i - \varphi(x_i, \gamma, \sigma)]^2} \tag{6.13}$$

where n is the number of training set; y_i is the actual value corresponding to x_i, and $\varphi(\cdot)$ is the nonlinear regression function determined by the LSSVM model. Therefore, the solution with a smaller RMSE has a larger probability of surviving in the successive generations.

Step 4: *Fitness evaluation*. Compute the fitness function value for each particle with (6.12) and (6.13), and find the best position p_{best} of each particle and the global best position g_{best} of all the particles in the current generation.

Step 5: *Weight evaluation, position and velocity update*. Evaluate the inertia weight value with Eq. (6.14) (Liu et al. 2011a, b), compute and update the position and velocity of each particle with Eqs. (6.15) and (6.16), respectively.

$$w(t) = w_{max} - \frac{w_{max} - w_{min}}{t_{max}} \times t \tag{6.14}$$

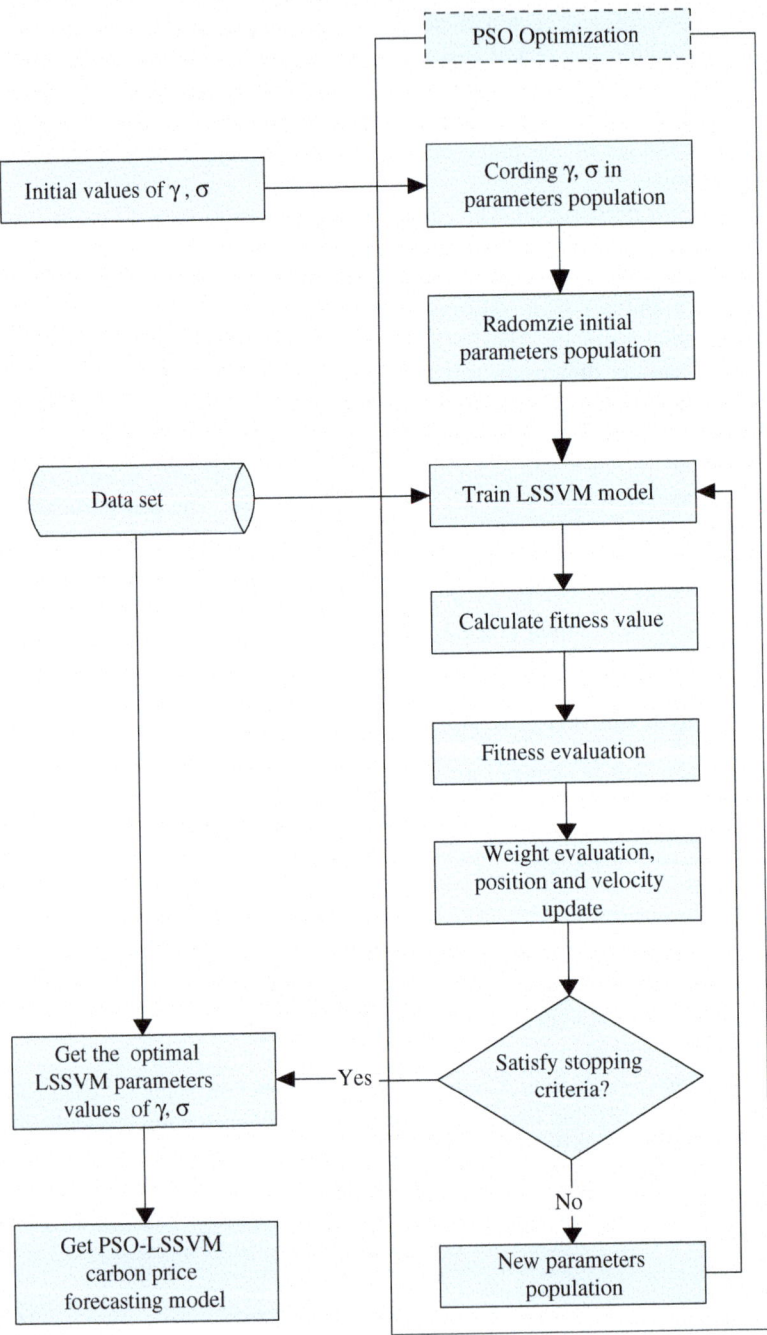

Fig. 6.1 PSO-LSSVM model

$$v_{id}(t+1) = \begin{cases} -v_{max}, v_{id} < -v_{max} \\ w(t) \cdot v_{id}(t) + c_1 \cdot r_1 \cdot [p_{id}(t) - x_{id}(t)] + c_2 \cdot r_2 \cdot [p_{gd}(t) - x_{id}(t)], -v_{max} \le v_{id} \le v_{max} \\ v_{max}, v_{id} > v_{max} \end{cases}$$

$$(6.15)$$

$$x_{id}(t+1) = \begin{cases} -p_{max}, x_{id} < -p_{max} \\ x_{id}(t) + v_{id}(t+1), -p_{max} \le x_{id} \le p_{max} \\ p_{max}, x_{id} > p_{max} \end{cases} \qquad (6.16)$$

where $1 \le i \le m$, $1 \le d \le D$, $v_{id}(t)$, and $x_{id}(t)$ are the current velocity and position of the particle i at iteration t, respectively, p_{id} is the best previous position of the particle i, and p_{gd} is the best position among all particles.

Step 6: *Stopping criteria check*. Stop the search process if the termination criterion such as maximum iteration is met, and output the optimal solution. At this point, the best LSSVM model is obtained. Otherwise, go to Step 7.

Step 7: $t = t + 1$, return to Step 4.

6.4   Forecasting of Carbon Prices

6.4.1   Data

As is well known, there are a great number of carbon prices in ECX. In this chapter, two main carbon future prices with maturity in December, 2016 (DEC16) and December, 2017 (DEC17) are chosen as experimental samples. The data of the two carbon prices used in this chapter are daily data, freely available from the ECX website.

For DEC16, we take daily data from November 27, 2012 to September 30, 2016, excluding public holidays, with a total of 985 observations. For DEC17, we take daily data from November 26, 2013 to September 30, 2016, excluding public holidays, with a total of 730 observations. For the convenience of LSSVM modeling, the data from November 27, 2012 to March 9, 2016 are used as the in-sample training sets in DEC16; the data from November 26, 2013 to April 4, 2016 are used as the in-sample training sets in DEC17(DEC16: 840 observations and DEC17: 600 observations), and the remainder are used as the out-of-sample testing sets (DEC16: 145 observations and DEC17: 130 observations), which are used to check the forecasting ability based on evaluation criteria. Figures 6.2 and 6.3 describe the curve of daily carbon prices for DEC16 and DEC17 in units of Euros/Ton, which shows that carbon prices have highly uncertain, nonlinear, dynamic and complicated characteristics. Therefore, it is difficult to accurately forecast the carbon prices.

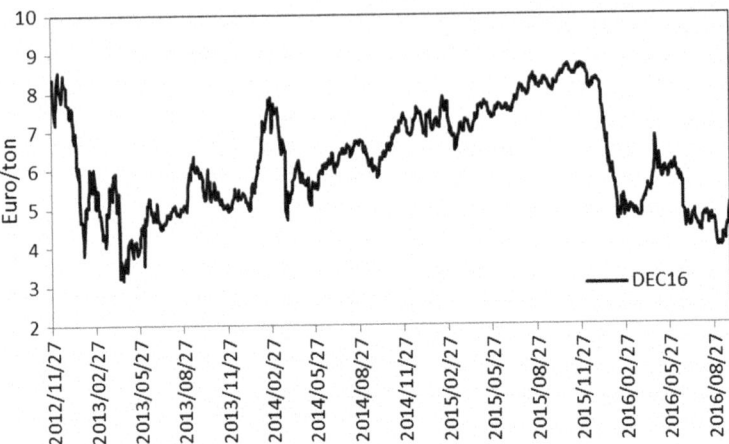

Fig. 6.2 ECX DEC16 carbon prices

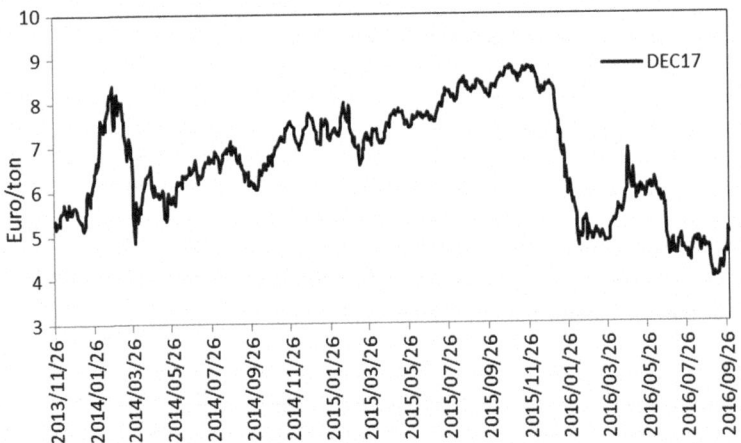

Fig. 6.3 ECX DEC17 carbon prices

6.4.2   Forecasting Evaluation Criteria

To measure the forecasting performance, we use two main criteria to evaluate the level prediction and directional forecasting. First, the root mean squared error (RMSE) is adopted as the evaluation criterion of level prediction, defined by formula (6.17) as:

$$RMSE = \sqrt{\frac{1}{n} \sum_{t=1}^{n} [\hat{x}(t) - x(t)]^2}. \qquad (6.17)$$

Second, the directional prediction statistic (D_{stat}) (Yu et al. 2010) is adopted as the evaluation criterion of directional forecasting, expressed as (6.18)

$$D_{\text{stat}} = \frac{1}{n} \sum_{t=1}^{n} \alpha_t \times 100\%,$$
(6.18)

where $x(t)$ is the actual value, $\hat{x}(t)$ is the predicted value, n is the number of predictions, $\alpha_t = 1$ if $[x(t+1) - x(t)][\hat{x}(t+1) - x(t)] \geq 0$, and $\alpha_t = 0$ otherwise.

However, there is an important question that whether the prediction accuracy of the model A is significantly better than that of the model B during the comparison of models. Thus, to evaluate models more objectively and robustly, the Diebold–Mariano test (DM; Diebold and Mariano 1995) is applied in evaluating the accuracy of level, while the Rate Test (RT) is applied in evaluating the accuracy of direction.

In this chapter, the mean square prediction error (MSPE) is chosen as the loss function. Thus, the DM statistic is defined as formula (6.19)

$$\text{DM} = \frac{\bar{d}}{\sqrt{V_{\bar{d}}/T}} \sim N(0,1), T \to \infty,$$
(6.19)

where $\bar{d} = \frac{1}{T} \sum_{t=1}^{T} [g(e_{te,t}) - g(e_{re,t})]$, $g(e_{te,t}) = \sum_{t=1}^{T} e_{te,t}^2$, $g(e_{re,t}) = \sum_{t=1}^{T} e_{re,t}^2$, $e_{te,t} = x_t - \hat{x}_{te,t}$, $e_{re,t} = x_t - \hat{x}_{re,t}$, $\hat{V}_{\bar{d}} = \gamma_0 + 2 \sum_{j=1}^{\infty} \gamma_j$, and $\gamma_j = \text{cov}(d_t, d_{t-j})$. $\hat{x}_{te,t}$ and $\hat{x}_{re,t}$ denote the forecasted values of x_t calculated using the test model (te) and reference model (re) at time t, respectively. A one-tailed test is generally employed in evaluating the DM statistic. In the DM test, the null hypothesis, i.e., the tested model is not worse than the reference model, is tested. Therefore, only if p is lower than a frequently used level of significance 0.05, we should reject it; otherwise, we should accept it.

The statistics of RT test is expressed as formula (6.20)

$$z_{\text{RT}} = \frac{p_A - p_B}{\sqrt{\frac{p_A(1-p_A)}{n}} + \sqrt{\frac{p_B(1-p_B)}{n}}} \sim N(0,1), n \to \infty$$
(6.20)

where p_A and p_B are respectively the accuracies of directional prediction of models A and B. The null hypothesis of RT test is that the accuracies of directional prediction of models A and B are the same. Using the two-sided test, when the absolute value of z_{RT} exceeds 1.96, the null hypothesis is rejected at the significance level of 5%.

In order to further ensure the robustness of the comparison results, the Superior predictive ability (SPA) test is also applied, where 8 different loss functions are selected to test the prediction accuracy of the models, that is $Li(i = 1, 2, \ldots, 8)$. $L1$ and $L2$ are mean squared error (MSE) and mean absolute error (MAE) respectively, which is commonly used in such judgments. Based on them, $L3$ and $L4$ take quadratic terms into consideration, meanwhile, $L5$ and $L6$ introduce the heteroscedasticity. Besides that, $L7$ is estimated by the Gauss likelihood function and $L8$ is similar with the value of M-Z regression. The specific definition of each loss function is as follows:

Proposed 8 loss function	
$L1 : \mathrm{MSE}_1 = n^{-1} \sum_{t=1}^{n} [x(t) - \hat{x}(t)]^2$	$L2 : \mathrm{MAE}_1 = n^{-1} \sum_{t=1}^{n} \|x(t) - \hat{x}(t)\|$
$L3 : \mathrm{MSE}_2 = n^{-1} \sum_{t=1}^{n} [x(t)^2 - \hat{x}(t)^2]^2$	$L4 : \mathrm{MAE}_2 = n^{-1} \sum_{t=1}^{n} \|x(t)^2 - \hat{x}(t)^2\|$
$L5 : \mathrm{HMSE} = n^{-1} \sum_{t=1}^{n} [1 - \frac{\hat{x}(t)}{x(t)}]^2$	$L6 : \mathrm{HMAE} = n^{-1} \sum_{t=1}^{n} \|1 - \frac{\hat{x}(t)}{x(t)}\|$
$L7 : \mathrm{QLIKE} = n^{-1} \sum_{t=1}^{n} [\ln(\hat{x}(t)) + \frac{x(t)}{\hat{x}(t)}]$	$L8 : R^2\mathrm{LOG} = n^{-1} \sum_{t=1}^{n} [\ln \frac{x(t)}{\hat{x}(t)}]^2$

$x(t)$ and $\hat{x}(t)$ are actual value and predicted value, respectively, while m is test period . It can be calculated from every prediction model $M_k, (k = 0, 1, \ldots, J)$ to get the 8 loss functions definited, which is denoted by $L_t^{i,k}$. Define M_0 as the base model for SPA test, for other J models $(k = 1, 2, \ldots, j)$, relative loss functions can be obtained as: $Y_t^{i,k} = L_t^{i,0} - L_t^{i,k}$. Null hypothesis H0 is: compared with the other model M_k, M_0 has the best performance in prediction. Construct the following statistic (6.21):

$$T^{SPA} = \max_{i,k} \frac{\bar{Y}_t^{i,k}}{\bar{\omega}_{kk}} \qquad (6.21)$$

$\bar{Y}_t^{i,k} = n^{-1} \sum_{t=1}^{n} Y_t^{i,k}$, meanwhile $\bar{\omega}_{kk}$ is the consistent estimation of standard error for $\bar{Y}_t^{i,k}$. The empirical distribution of T^{SPA} can be obtained from the process of Bootstrap. The parameters of Bootstrap are set as: 10,000 and 0.5 for the resamples and the dependence, respectively. The higher P-value (closer to 1) of SPA test, the more the null hypothesis can not be rejected, which means a higher prediction accuracy of the base model.

A good statistical accuracy does not always mean a good trading performance. For investors, they usually care more about a model's practicability in trading. In this section, inspired by Sermpinis et al. (2016), we design a strategy to test the trading performance as an investor chooses to buy or sell(or stay watching) carbon when the forecasted return is above or below(or equal) zero at the current carbon price respectively. We use the Annualized return, Annualized volatility, and Information ratio to evaluate the trading performance. They defined as:

Measures of trading performance	
Daily return	$R_t = \frac{P_t - P_{t-1}}{P_{t-1}}$
Annualized return	$R^A = 252 \times \frac{1}{k} \sum_{t=1}^{k} R_t$
Annualized volatility	$\sigma^A = \sqrt{252} \times \sqrt{\frac{1}{k-1} \sum_{t=1}^{k} (R_t - R_m)^2}$
Information ratio	$IR = \frac{R^A}{\sigma^A}$

where, P_t is the daily price of carbon futures at day t; k is the number of test set, and R_m is mean of R_t.

For comparing the prediction capacity of the proposed ARIMA and LSSVM hybrid forecasting models with other widely used forecasting approaches, we employ the single ARIMA and LSSVM models as benchmark models. Moreover, the hybrid ARIMA and LSSVM models and the COMBINED model (weighting predictive value of ARIMA and LSSVM models), are also used to predict carbon prices for the purpose of comparisons.

6.4.3   Parameters Determination of Three Models

(1) ARIMA model

The DEC16 and DEC17 carbon price data are varied with daily change, and show characteristics of tendency and instability. The ARIMA model is fitted to a stationary time series, and the daily data require regular differencing to become stationary. In this chapter, the ARIMA model is implemented via the Eviews statistical software package, produced by Quantitative Micro Software Corporation. The Akanke Information Criteria (AIC) is used to identify the best model. By trial and error in the two single models, both the optimal models generated from the DEC16 and DEC17 carbon prices are ARIMA (2, 1, 2), i.e., $p = 2$, $d = 1$ and $q = 2$, and the final parameter estimation results are shown in Tables 6.1 and 6.2 for DEC 16 and DEC17, respectively.

Table 6.1 Parameters estimation results of the ARIMA for DEC16

Variable	Coefficient	Std. error	t-statistic	Prob.
C	−0.003597	0.006429	−0.559527	0.576
AR(1)	−0.389558	0.006515	−59.79822	0.0000
AR(2)	−0.979514	0.006501	−150.6744	0.0000
MA(1)	0.420524	0.003234	130.0227	0.0000
MA(2)	0.994357	0.002584	384.8171	0.0000
R-squared	0.035557	Mean dependent variable		−0.00368
Adjusted	0.03092	S.D. dependent variable		0.185425
S.E. of regression	0.182536	Akaike info criterion		−0.55778
Sum squared resid	27.72182	Schwarz criterion		−0.529525
Log likelihood	238.4309	Hannan–Quinn criterion		−0.546949
F-statistic	7.668549	Durbin–Watson statistic		1.893166
Prob (F-statistic)	0.000005			

Table 6.2 Parameters estimation results of the ARIMA for DEC17

Variable	Coefficient	Std. error	t-statistic	Prob.
C	−0.000612	0.005992	−0.102103	0.9187
AR(1)	−1.024049	0.1776	−5.766024	0.0000
AR(2)	−0.749565	0.137297	−5.459441	0.0000
MA(1)	1.092025	0.174885	6.244248	0.0000
MA(2)	0.771026	0.137055	5.625661	0.0000
R-squared	0.013472	Mean dependent variable		−0.000646
Adjusted	0.007769	S.D. dependent variable		0.153867
S.E. of regression	0.153268	Akaike info criterion		−0.90611
Sum squared resid	16.25583	Schwarz criterion		−0.873493
Log likelihood	320.7792	Hannan–Quinn criterion		−0.893499
F-statistic	2.362397	Durbin–Watson statistic		2.048449
Prob (F-statistic)	0.05186			

(2) *LSSVM model*

For the LSSVM model, we use the LSSVMlab software package toolbox developed by Suykens et al. for MATLAB 2012b platform to build the LSSVM models. The RBF function is chosen as the kernel function of the LSSVM models. There is no standard procedure for determining the parameters, γ and σ, for the LSSVM model. Thus, we apply PSO to synchronously optimize both of them. The inputs of LSSVM models are determined in the same manner as the proposed hybrid models. PSO is applied to seek and obtain the LSSVM optimal parameter sets when the RMSE is at its minimum. The searching process of optimal parameters is operated with real code, 100 initial particles, 20 generations, $\gamma \in (0.01, 10000)$, $\sigma \in (0.1, 100)$, $c_1 = c_2 = 2$, $w \in [0.1, 0.9]$, $p_{max} = 0.05$, and $v_{max} = 50$, as well as fitness function $F_{fitness} = RMSE(\gamma, \sigma)$.

(3) *Hybrid models*

In this investigation, we propose three hybrid models fundamentally derived from LSSVM model. Therefore, the determining process for parameters is similar to the above-mentioned LSSVM model. The obtained optimal parameters for each hybrid model are as shown in Table 6.3.

Table 6.3 Optimal parameters for each hybrid model

Model	DEC16		DEC17	
	γ	σ	γ	σ
ARIMALSSVM1	0.1	6.6392	0.1	316.2278
ARIMALSSVM2	3519.6789	6.9878	1125.7778	10
ARIMALSSVM3	10,000	4.0056	4864.0270	8.4916

6.4.4   Statistical Performance

Firstly, the two main carbon price datasets, DEC16 and DEC17, are applied to test these three hybrid models, which integrate the ARIMA and LSSVM models and are built with different inputs. In the meantime, the RMSE, D_{stat}, DM, RT and SPA calculated on the testing sets are applied to evaluate the forecasting performance of these three hybrid models. Only if a model has a smaller RMSE, DM, RT and/or larger D_{stat}, SPA, will it signify a better model. After the optimal parameter sets of LSSVM are gained by PSO, these forecasting models are built. The out-of-sample forecasting results are reported in Tables 6.4, 6.5, 6.6, 6.7, 6.8, 6.9, 6.10 and 6.11. As in Table 6.4, the ARIMALSSVM2 model (input nodes Y_{t-1}, Y_t and residual value ε_t) has the best out-of-sample forecasting accuracy due to the smallest RMSE and the largest D_{stat} among three hybrid models.

Table 6.4 The out-of-sample forecasting comparisons of different hybrid models

Model	DEC16		DEC17	
	RMSE	D_{stat}	RMSE	D_{stat}
ARIMLSSVM1	0.1501	57.93	0.1584	66.15
ARIMALSSVM2	**0.0346**	**91.03**	**0.0386**	**96.15**
ARIMALSSVM3	0.1561	57.93	0.1615	56.15

Table 6.5 The out-of-sample forecasting comparison of different models

Model	DEC16		DEC17	
	RMSE	D_{stat}	RMSE	D_{stat}
ARIMALSSVM2	0.0345	91.03	0.0309	96.15
COMBINED	0.1475	66.70	0.1549	64.62
LSSVM	0.1474	61.38	0.1577	66.92
ARIMA	0.1483	59.31	0.1589	54.62

Table 6.6 The out-of-sample forecasting comparisons of DM test for DEC16

Test model	Reference model				
	ARIMA	LSSVM	ARIMALSSVM1	ARIMALSSVM2	ARIMALSSVM3
LSSVM	0.2521				
ARIMALSSVM1	0.3519	0.3798			
ARIMALSSVM2	0.0000*	0.0000*	0.0000*		
ARIMALSSVM3	0.2751	0.0569	0.1883	0.0000*	
COMBINED	0.0492*	0.8320	0.1653	0.0000*	0.0487*

Note This table reports the *P*-value of DM test. *Denotes that the null hypothesis is rejected at the significant level of 5%

Table 6.7 The out-of-sample forecasting comparisons of DM test for DEC17

Test model	Reference model				
	ARIMA	LSSVM	ARIMALSSVM1	ARIMALSSVM2	ARIMALSSVM3
LSSVM	0.6453				
ARIMALSSVM1	0.0152*	0.8655			
ARIMALSSVM2	0.0000*	0.0000*	0.0000*		
ARIMALSSVM3	0.4709	0.1235	0.3556	0.0000*	
COMBINED	0.4001	0.9649	0.7503	0.0000*	0.1345

Note This table reports the *P*-value of DM test. *Denotes that the null hypothesis is rejected at the significant level of 5%

Table 6.8 The out-of-sample forecasting comparisons of RT test in DEC16

Test model	Reference model				
	ARIMA	LSSVM	ARIMALSSVM1	ARIMALSSVM2	ARIMALSSVM3
LSSVM	0.3398				
ARIMALSSVM1	0.8127	0.0000*			
ARIMALSSVM2	0.0000*	0.0000*	0.0000*		
ARIMALSSVM3	0.7220	0.0000*	0.0000*	0.0000*	
COMBINED	0.1181	0.0000*	0.0000*	0.0000*	0.0000*

Note This table reports the *P*-value of RT test. *Denotes that the null hypothesis is rejected at the significant level of 5%

Table 6.9 The out-of-sample forecasting comparisons of RT test in DEC17

Test model	Reference model				
	ARIMA	LSSVM	ARIMALSSVM1	ARIMALSSVM2	ARIMALSSVM3
LSSVM	0.2033				
ARIMALSSVM1	0.7936	0.1253			
ARIMALSSVM2	0.0000*	0.0000*	0.0000*		
ARIMALSSVM3	0.1282	0.8026	0.0749	0.0000*	
COMBINED	0.5204	0.5282	0.3660	0.0000*	0.3786

Note This table reports the *P*-value of RT test. *Denotes that the null hypothesis is rejected at the significant level of 5%

Table 6.10 The out-of-sample forecasting comparisons of SPA test of each prediction models

Models	DEC16							
	MSE1	MSE2	MAE1	MAE2	HMSE	HMAE	QLIKE	RLOG
ARIMA	0.0001	0.0008	0.0000	0.0000	0.0000	0.0000	0.0000	0.0000
LSSVM	0.0001	0.0003	0.0000	0.0000	0.0000	0.0000	0.0000	0.0000
ARIMALSSVM1	0.0001	0.0002	0.0000	0.0000	0.0000	0.0000	0.0000	0.3292*
ARIMALSSVM2	0.5626*	0.5560*	0.5116*	0.5184*	0.6076*	0.5175*	0.6007*	0.6908*
ARIMALSSVM3	0.0002	0.0008	0.0000	0.0000	0.0002	0.0000	0.0001	0.0001
COMBINED	0.0001	0.0003	0.0000	0.0000	0.0000	0.0000	0.0000	0.0000

Note *Denotes that the null hypothesis is rejected at the significant level of 10%

Table 6.11 The out-of-sample forecasting comparisons of SPA test of each prediction models

Models	DEC17							
	MSE1	MSE2	MAE1	MAE2	HMSE	HMAE	QLIKE	RLOG
ARIMA	0.0001	0.0006	0	0	0	0	0	0
LSSVM	0.0002	0.0004	0	0	0	0	0	0
ARIMALSSVM1	0.0001	0.8358*	0	0	0	0	0	0
ARIMALSSVM2	0.6352*	0.1642*	0.5708*	0.5657*	0.5581*	0.5454*	0.5567*	0.5583*
ARIMALSSVM3	0.0001	0.0008	0	0	0.0002	0	0.0001	0.0001
COMBINED	0	0.0006	0	0	0	0	0	0

Note *Denotes that the null hypothesis is rejected at the significant level of 10%

Secondly, we employ the single ARIMA and LSSVM models as benchmark models. The COMBINED model is also used to predict carbon price for the purpose of comparison. The LSSVM is established with the LSSVMlab 1.5 Toolbox of the Matlab software package, produced by Suykens et al. LSSVM has also one input node (Y_{t-2}, Y_{t-1}) and one output (Y_t), and RBF kernel function is selected as the kernel function. At the same time, these models are trained by PSO with the same parameters setting. The out-of-sample forecasting results are presented in Table 6.5.

Moreover, we compare the level prediction abilities of the hybrid model (ARIMALSSVM2), the single models (ARIMA and LSSVM models), and the COMBINED model in term of the RMSE indicator. The ARIMALSSVM2 model performs best, ARIMALSSVM3 performs the worst in all cases. The level prediction abilities of ARIMALSSVM1 model and ARIMALSSVM3 model are lower than ARIMALSSVM2. The COMBINED model and ARIMALSSVM2 was better than the single model (LSSVM and ARIMA) in term of the RMSE. In addition, we also find that the single LSSVM model outperforms the single ARIMA model, for the LSSVM shows a bigger advantage than ARIMA in term of the RMSE indicator. One finding is derived from the DM test results, that the hybrid ARIMALSSVM2 model remarkably outperforms than single models and the COMBINED model at the significance level of 5%. There is no obvious difference between ARIMA, LSSVM, and COMBINED models in level prediction abilities. From the SPA test, the hybrid ARIMALSSVM2 model remarkably outperforms than single models and COMBINED model at the highest probability. The other models, except ARIMALSSVM2, show low probability to be the best model in the term of different loss function, but have no obvious disadvantage than the others. Interestingly, the ARIMALSSVM2 model performs much better than the single models in all the cases. One possible reason could be that the hybrid strategy impacts the forecasting performance. In addition, we also find that the single LSSVM model outperforms the single ARIMA model. The possible explanations could be twofold; on one hand, the LSSVM model is a class of nonlinear forecasting model, which can capture nonlinear patterns hidden in the carbon prices, while the single ARIMA model is a class of linear models, which is not suitable for predicting the nonlinear carbon prices. Therefore, for the out-of-sample error comparisons of DEC16 and DEC17, the hybrid ARIMALSSVM2 model is superior to the single ARIMA and LSSVM models, and COMBINED model.

Furthermore, we evaluate the directional forecasting abilities of the hybrid model (ARIMALSSVM2), the single models (ARIMA and LSSVM models), and the COMBINED model in terms of the D_{stat} indicator, which is more important for business investors than the RMSE indicator (Yu et al. 2010). The larger D_{stat}, the greater the directional forecasting ability. For out-of-sample error comparisons of DEC16 and DEC17, the D_{stat} of the ARIMALSSVM2 models are the largest, at 91.03 and 96.15%, respectively, followed by COMBINED, LSSVM and ARIMA models. The COMBINED model has advantage than LSSVM model and ARIMA model in the term of D_{stat} indicator. The LSSVM model performs a little advantage than ARIMA in the term of D_{stat} indicator. And the RT test reveals the similar results as the DM test, i.e., the hybrid ARIMALSSVM2 model can obtain high accuracy of directional prediction at the confidence level of 5% for all carbon prices. There is no obvious advantage between ARIMA, LSSVM, and COMBINED models in the directional forecasting ability. The main reason of this phenomenon is the poor directional change detectability of the single models in comparison to the increased overall forecasting accuracy of the hybrid models. Interestingly, same as the RMSE indicator, the single LSSVM model performs better than the single ARIMA model in terms of the D_{stat} indicator. As has already been discussed, the likely explanation is that the carbon price has highly nonlinear and complex characteristics. Therefore, the hybrid ARIMALSSVM2 model has an excellent ability to forecast turning points, which shows that the hybrid ARIMALSSVM2 model can yield significant forecast improvements. The gains in forecast accuracy may result from the ability of the integrated ARIMA, PSO, and LSSVM models to capture both linear and nonlinear patterns existing in the carbon price.

6.4.5   Trading Performance

In terms of the trading performance, shown in Table 6.12, the proposed ARIMALSSVM2 model produces the best trading performances in all the carbon prices for its highest annualized return and smallest annualized volatility. The ARIMA model performs lowest annualized return and biggest annualized volatility. The LSSVM model perform better than ARIMA model for its higher Annualized return and smaller Annualized volatility. The COMBINED model show lower disadvantage than the LSSVM model and higher advantage than the ARIMA model. This implies that our proposed ARIMALSSVM2 model is capable of achieving good trading gains.

In brief, according to the experimental results of carbon price forecasting for ECX DEC16 and DEC17 presented in this chapter, we can draw some conclusions: (1) The experimental results show that the ARIMALSSVM2 model can exceed the single ARIMA and LSSVM models, as well as the hybrid ARIMALSSVM1, ARIMALSSVM3, and COMBINED models, for the testing sets of the two main carbon future prices in terms of level prediction, directional prediction, and trading

Table 6.12 The out-of-sample forecasting comparisons of different models

Trading performance	DEC16			DEC17		
	Annualized return (%)	Annualized volatility (%)	Information ratio	Annualized return (%)	Annualized volatility (%)	Information ratio
ARIMA	35.4	43.49	0.81	−68.01	39.56	−1.72
LSSVM	74.64	41.51	1.8	67.95	45.31	1.50
ARIMALSSVM1	58.58	44.08	1.33	−13.80	41.60	−0.33
ARIMALSSVM2	503.73	31.42	16.03	554.64	30.87	17.97
ARIMALSSVM3	48.04	43.13	1.11	40.69	44.28	0.92
COMBINED	57.22	38.93	1.47	38.04	43.64	0.87

performances; (2) Not all the combined/hybrid models are superior to the single models, which indicates that the combined /hybrid principles cannot effectively improve the prediction performance in every situation. (3) The ARIMALSSVM2 hybrid forecasting model can significantly improve forecasting accuracy; namely, the ARIMALSSVM2 model outperforms other forecasting models based on RMSE, D_{stat}, DM, RT, SPA and trading performance which results in the fourth conclusion. (4) The hybrid ARIMALSSVM2 model seems suitable for global carbon prices forecasting.

6.5   Conclusions

In this chapter, we have presented several novel hybrid models incorporating ARIMA with LSSVM models for carbon price forecasting. The ARIMA model is applied to capture the linear patterns hidden in the carbon prices, whilst the LSSVM is used to capture the nonlinear patterns existing in the carbon prices, resulting in a hybrid methodology which can improve forecasting accuracy. The hybrid models mainly consist in the evolutionary training of an LSSVM using the PSO algorithm. Once the optimal model parameters have been determined by the PSO, we can efficiently predict the carbon prices of the future. This chapter has evaluated and compared the hybrid ARIMALSSVM2 model with the single ARIMA and LSSVM models, as well as the hybrid ARIMALSSVM1, ARIMALSSVM3, and COMBINED models, using RMSE, D_{stat}, DM, RT, SPA, and trading performance as the criteria using two well-known carbon price datasets from the ECX market.

The empirical results show that the hybrid model (ARIMALSSVM2) can produce the lowest RMSE, DM, RT and the highest D_{stat}, SPA and the best trading performance in the carbon price data sets. It exceeds the single ARIMA and LSSVM models, as well as the hybrid ARIMALSSVM1, ARIMALSSVM3, and COMBINED model. However, not all the combined/hybrid models are superior to the single models in terms of both level prediction and directional prediction. That is to say, compared with the single models, not all the combined/hybrid models can consistently achieve superior predictive performances, which has also been verified by the previous studies (Khashei and Bijari 2011; Taskaya and Casey 2005; Yu et al. 2005). Most of the single models and hybrid models evaluated show poor ability to detect directional change, as evident from the RT indicator. The hybrid model D_{stat} (ARIMALSSVM2) can effectively overcome this problem. Besides, excellent turning points detectability, the hybrid model (ARIMALSSVM2) can achieve superior forecasting performances and show good results. Therefore, the proposed hybrid model (ARIMALSSVM2) seems suitable for forecasting the highly nonlinear and complex carbon price.

References

Bataller MM, Pardo A, Valor E (2007) CO_2 prices energy and weather. Energy J 28(3):73–92

Beat H (2010) Allowance price drivers in the first phase of the EUETS. J Environ Econ Manag 59:43–56

Cagdas HA, Erol E, Cem K (2009) Forecasting nonlinear time series with a hybrid methodology. Appl Math Lett 22:1467–1470

Chen KY, Wang CH (2007) A hybrid SARIMA and support vector machines in forecasting the production values of the machinery industry in Taiwan. Expert Syst Appl 32:254–264

Diebold FX, Mariano RS (1995) Comparing predictive accuracy. J Bus Econ Stat 13(3):253–263

Firat M, Gungor M (2009) Generalized regression neural networks and feed forward neural networks for prediction of scour depth around bridge piers. Adv Eng Softw 40(8):731–737

Ioannou K, Arabatzis G, Lefakis P (2009) Predicting the prices of forest energy resources with the use of Artificial Neural Networks (ANNs): the case of conifer fuel wood in Greece. J Environ Prot Ecol 10(3):678–694

Jain A, Kumar AM (2007) Hybrid neural network models for hydrologic time series forecasting. Appl Soft Comput 7:585–592

Kavaklioglu K (2011) Modeling and prediction of Turkey's electricity consumption using Support Vector Regression. Appl Energy 88:368–375

Kennedy J, Eberhart RC (1995) Particle swarm optimization. In: Proceedings of the IEEE Conference on Neural Networks, vol 4. Piscataway, Perth, pp 1942–1948

Keppler JH, Bataller MM (2010) Causalities between CO_2, electricity, and other energy variables during phase I and phase II of the EU ETS. Energy Policy 38:3329–3341

Khashei M, Bijari M (2011) A novel hybridization of artificial neural networks and ARIMA models for time series forecasting. Appl Soft Comput 11:2664–2675

Kong F, Wu XJ (2008) Time series forecasting model with error correction by structure adaptive support vector machine. In: Proceedings of international conference on computer science and software engineering. pp 1067–1070

Koutroumanidis T, Ioannou K, Arabatzis G (2009) Predicting fuel wood prices in Greece with the use of ARIMA models, artificial neural networks and a hybrid ARIMA-ANN model. Energy Policy 37:3627–3634

Liu LX, Zhuang YQ, Liu XY (2011a) Tax forecasting theory and model based on SVM optimized by PSO. Expert Syst Appl 38:116–120

Liu XY, Shao C, Ma HF, Liu RX (2011b) Optimal earth pressure balance control for shield tunneling based on LS-SVM and PSO. Autom Constr 20:321–327

Omer FD (2009) A hybrid neural network and ARIMA model for water quality time series prediction. Eng Appl Artif Intell. doi:10.1016/j.engappai.2009.09.015

Pai PF, Lin CS (2005) A hybrid ARIMA and support vector machines model in stock price forecasting. Omega 33:497–505

Seifert J, Marliese UH, Michael W (2008) Dynamic behavior of CO_2 spot prices. J Environ Econ Manag 56:180–194

Sermpinis G, Stasinakis C, Rosillo R, Fuente D (2016) European exchange trading funds trading with locally weighted support vector regression. Eur J Oper Res 258:372–384

Shafie-khah M, Parsa Moghaddam M, Sheikh-El-Eslami MK (2011) Price forecasting of day-ahead electricity markets using a hybrid forecast method. Energy Convers Manag 52:2165–2169

Smola AJ (1998) Learning with kernels. Ph.D. thesis, Department of Computer Science, Technical University Berlin, Germany

Suykenns JAK, Vandewalle J (1999) Least squares support vector machine. Neural Process Lett 9 (3):293–300

Taskaya T, Casey MC (2005) A comparative study of autoregressive neural network hybrids. Neural Netw 18:781–789

Tseng FM, Yu HC, Tzeng GH (2002) Combining neural network model with seasonal time series ARIMA model. Technol Forecast Soc Chang 69:71–87

Vapnik VN (1995) The nature of statistical learning theory. Springer, New York

Vapnik VN, Golowich S, Smola A (1997) Support vector method for function approximation, regression estimation and signal processing. Advance in neural information processing system, vol 9. MIT Press, Cambridge, pp 281–287

Wei YM, Wang K, Feng ZH et al (2010) Carbon finance and carbon market: models and empirical analysis. Science Press, Beijing

Yu L, Wang SY, Lai KK (2005) A novel nonlinear ensemble forecasting model incorporating GLAR and ANN for foreign exchange rates. Comput Oper Res 32:2523–2541

Yu L, Wang SY, Lai KK et al (2010) A multiscale neural network learning paradigm for financial crisis forecasting. Neurocomputing 73:716–725

Zhang GP (2003) Time series forecasting using a hybrid ARIMA and neural network model. Neurocomputing 50:159–175

Zhang YJ, Wei YM (2010) An overview of current research on EU ETS: evidence from its operating mechanism and economic effect. Appl Energy 87(6):1804–1814

Zhang Y, Wu L (2009) Stock market prediction of S&P 500 via combination of improved BCO approach and BP neural network. Expert Syst Appl 36(5):8849–8854

Chapter 7
Carbon Price Forecasting Using a Parameters Simultaneous Optimized Least Squares Support Vector Machine with Uniform Design

Abstract This chapter augments the least squares support vector machine (LSSVM) approach of forecasting carbon prices by adding the uniform design (UD) feature. Compared with the particle swarm optimization (PSO) feature, uniform design displays optimization efficiency advantages in complex (low/high) carbon price environments.

7.1 Introduction

As an efficient way for controlling the cost of tackling climate change, carbon market has been paid more and more attention in theoretical and practical fields. During the past few years, global carbon market, represented by the EU ETS, has witnessed a rapid development. However, carbon price fluctuates violently, which remarkably impacts the emission reduction performance and carbon market value. Predicting carbon price accurately is not only benefit for understanding the patterns of carbon price fluctuations and constructing an efficient mechanism of carbon price stabilization, but also is conductive to make investors avoid carbon market risks and increase value of carbon assets. For this reason, carbon price forecasting has become one of the urgent topics in the fields of energy and climate change (Zhang and Wei 2010).

In the early stage of carbon market, carbon price is mainly qualitatively forecasted (Reilly and Paltsev 2005; Kanen 2006). During recent years, more and more quantitative methods have been introduced into carbon price forecasting, among which econometric models and statistical models including GARCH (Paolella and Taschini 2008; Byun and Cho 2013), MS-AR-GARCH (Benz and Truck 2009), FAVAR (Chevallier 2010), FIAPGARCH (Conrad et al. 2012), HAR-RV (Chevallier and Sevi 2011), a nonparametric statistics method (Chevallier 2011), the dynamic model averaging method (Koop and Tole 2013) are most commonly used. However, as carbon price change is highly nonlinear and nonstationary while the traditional econometric models and statistical models are constructed in the

Special thanks to Xuetao Shi, Dong Han, Ping Wang and Ying-Ming Wei for actively counseling to the writing of Chap. 7.

assumption that the data are stationary and linear, those models can fail to efficiently deal with the nonstationary and nonlinear patterns concealed in the carbon price, and therefore are hard to achieve the accurate forecasting results of carbon price.

During the past few years, aiming at the limitations of the traditional econometric models and statistical models, nonlinear artificial intelligent methods, represented by artificial neural network (ANN) and support vector machine (SVM) (Zhu 2012), have been adopted in carbon price forecasting. As being established on statistical learning theory, SVM can show higher predication accuracy than the traditional statistical models, econometric models, and ANN models. Least squares support vector machine (LSSVM), as a modified SVM through replacing inequality constraints with equality ones, can avoid to solve quadratic programming, and therefore can reduce the computation complexity and accelerate the solution speed. Owing to these advantages, LSSVM has been applied in carbon price forecasting in recent years, and achieved better prediction effects than the traditional models (Zhu and Wei 2013). Although LSSVM can present a favorable nonlinear prediction and modeling capacity, and obtain a higher forecasting accuracy than the traditional methods, phase space reconstruction (PSR) and LSSVM parameters selection are complex and difficult. The main reasons are: (i) The parameters of LSSVM model directly decide its learning and prediction capacity, while there is no effective approach for the LSSVM parameters selection so far. (ii) Carbon price is usually nonlinear, nonstationary and chaotic (Fan et al. 2015). It is hard to determine the optimal delay time (τ) and embedding dimension (m) in the PSR. Traditionally, separate optimization and alternative optimization are used widely to determine the PSR and model parameters of LSSVM. Separate optimization refers to respectively determine the PSR and parameters of LSSVM. Alternative optimization refers to take turns to determine the PSR and parameters of LSSVM, first determine one then determine the other. For the PSR, trial-and-error, autocorrelation function, average mutual information, Cao method, τ-m and C–C methods (Kim et al. 1999; Fraser 1989; Maguire et al. 1998; Cao 1997) are widely mainly applied, while for parameters of LSSVM, trial-and-error, gradient descent algorithm, n-fold cross validation, grid search method, genetic algorithm, and particle swarm optimization (PSO) (Zhou et al. 2009; Zhang et al. 2013; Silva et al. 2015) are mainly utilized. In fact, during the modeling of LSSVM for carbon price forecasting, optimization of PSR, and LSSVM parameters are interdependent. That is to say, either separate optimization or alternative optimization cannot be verified without the other. The optimization of τ, m and parameters of LSSVM cannot be separate or alternative. Either separate optimization or alternative optimization breaks the internal links between them and cannot ensure that the both are optimized simultaneously. As a result, the prediction results are unsatisfactory. Therefore, accurate forecast of carbon price, as a challenging work, has to simultaneously optimize the PSR and parameters of LSSVM, so as to obtain comprehensively optimal prediction results of carbon price. Actually, the optimization of parameter combination is an optimization of multiple factors and multiple levels, which is very time-consuming. As an experimental optimization method, in essence, uniform design (UD) can take full use of the distribution uniformity of experimental points, and obtain the maximum information through minimum experiments. So, we introduce uniform design into the parameter optimization of PSR and LSSVM parameters to translate the large sample problem of parameter optimization into a small sample problem, so as to improve the optimization efficiency at the same time ensuring the high optimization effect.

The aim of this chapter is to overcome the limitations of separate optimization and alternative optimization of PSR and parameters of LSSVM, and explore a novel parameters simultaneous optimization method (UD-LSSVM) for PSR and LSSVM with uniform design for carbon price forecasting, so as to furthest improve the forecasting accuracy and modeling efficiency. The innovations lie in the following two aspects: on the one hand, a uniform design-based parameters simultaneous optimization approach for PSR and parameters of LSSVM is constructed for carbon price forecasting. The parameters simultaneous optimization of PSR and LSSVM is a large sample combinatorial optimization of multiple factors and multiple levels in essence, which is transformed into a small sample combinatorial optimization using uniform design so as to enhance the optimization efficiency. All the parameters are simultaneously optimized using the self-invoking LSSVM to obtain the optimal parameters. On the other hand, the proposed method is verified by forecasting two carbon futures prices with different maturities under the EU ETS. The empirical results obtained that compared with PSO, the proposed method can significantly improve the modeling efficiency at the same time ensuring a high prediction accuracy.

7.2 Methodology

7.2.1 Parameter Selection of a LSSVM Predictor

In this chapter, we select LSSVM as the predictor of carbon price. LSSVM is proved to be a power learning method for the regression problems. However, in the modeling of LSSVM, how to select the kernel function can directly determine its performance. Each kernel function has its own applicative data distribution patterns. That is to say, for the same data, the performances of LSSVM models with different kernel functions are different. A large number of studies have shown that the Radial Basis kernel function (RBF), $K(x, y) = \exp(\frac{-\|x-y\|^2}{2\sigma^2})$, has a strong nonlinear mapping capacity. For the data with no prior knowledge, the performance of LSSVM model with RBF kernel function is generally better than that of LSSVM models with the other kernel functions (Liu et al. 2011). In this chapter, RBF kernel function is applied as the kernel function of LSSVM predictor. Therefore, parameter selection for the LSSVM predictor includes as follows.

(1) *Phase space reconstruction (PSR)*. PSR is a basis for predicting carbon price using the LSSVM model. PSR involves the selections of optimal delay time (τ) and embedding dimension (m). Too small m cannot display the fine structure of chaotic system for carbon price, while too large m will complicate the computation and therefore can cause noise. Likewise, if τ is too small, the adjacent delay

coordinate elements differ slightly in the phase space and therefore can lead to information redundancy, whereas with a overlarge τ, the adjacent delay coordinate elements are not associated, which can result into information lose and thus can fold signal trajectories (Albano et al. 2002). Therefore, the accurate selections of m and τ can directly influence the forecasting results of carbon price.

(2) *Penalty factor γ of LSSVM and parameter σ of RBF kernel function.* γ can determine the complexity of the LSSVM model and the penalty of fitting errors. σ can represent the structure of high-dimensional space, reflect the characteristics of training data, and influence the generalization of LSSVM model. The larger γ, the closer the fitting value of training data is to the corresponding actual value, while over-fitting is likely to happen. A less γ reduces the complexity of LSSVM model, whereas insufficient learning probably occurs. σ is too small, over-fitting is likely to occur with a weak generalization. While σ is too large, there is possibly insufficient learning. γ and σ^2 are generally in the ranges of [1, 100,000] and [0.1, 50,000], respectively.

(3) *Sliding time-window w.* As the carbon price varies over time, with the acquisition of new input and output data, the state of carbon price changes continuously. So, to make the established LSSVM model reflect the current state of carbon price accurately, the LSSVM model should be described using new data, while old data which are less related to the current state of carbon price should be ignored or hold a small weight. Furthermore, based on the principle that the newer the data to the forecasting time point, the larger influences on the forecasting results, and vice versa, a modeling data range, namely sliding time-window, slides with time needs to be established to make the LSSVM model introduce by waves the newly acquired carbon price information, so as to reveal the time structural characteristics of carbon price data.

In brief, to obtain a carbon price forecasting model with optimal performance, the parameters including m, τ, γ, σ, and w have to be optimized in the modeling of a LSSVM predictor. All these parameters are highly associated in fact. Obviously, it is a combinational optimization problem. When the exhaustive search method is used to optimize the whole combinations for each parameter and each level, the optimization needs to be performed for lots of times, which can cause a remarkable amount of computation, and even sometimes the optimization cannot be realized. To reduce the optimization times, in traditional methods, the optimal level of each parameter is found out successively, thus transferring the optimization of multiple parameters to the optimization of single parameter. However, in the traditional methods, the links among various parameters are ignored. Therefore, the combination of the optimal levels of all the parameters is possibly not the optimal parameter combination. Actually, the optimization of parameter combination is an optimization of multiple factors and multiple levels, which is very time-consuming. To shorten the optimization time, uniform design is introduced into the parameter optimization, and self-invoking LSSVM is adopted to translate the large sample problem of parameter optimization into a small sample problem, so as to accelerate the optimization speed.

7.2.2   Uniform Design for Parameter Selection of a LSSVM Predictor (UD-LSSVM)

Uniform design can take full use of the distribution uniformity of experimental points, and obtain the maximum information through minimum experiments, which can dramatically decrease the optimization time. Therefore, uniform design is particularly suitable for an experiment of multiple factors and multiple levels (Fang 1994). In this research, uniform design is applied in the initial design of model parameters for carbon price forecasting, and then self-invoking learning is conducted using LSSVM to acquire the optimal parameters, which are used for carbon price forecasting. The modeling process of UD-LSSVM algorithm is demonstrated in Fig. 7.1, and the concrete procedures are as follows:

Step 1: *Presetting the initial ranges for model parameters.* The upper and lower limits of all the parameters to be optimized are set as $[m_{\min}, m_{\max}]$, $[\tau_{\min}, \tau_{\max}]$, $[\gamma_{\min}, \gamma_{\max}]$, $[\sigma_{\min}, \sigma_{\max}]$, and $[w_{\min}, w_{\max}]$ referring to previous researches.

Step 2: *Generating the initial parameter combinations using uniform design.* Based on the upper and lower limits as well as the step size, the initial parameter combinations are generated through uniform design.

Step 3: *PSR and LSSVM predictor establishment.* First, carbon price data are reconstructed by applying the m and τ of the initial parameter combination in the PSR. Afterwards, γ and σ are adopted as the parameters of LSSVM to train and forecast the reconstructed carbon price. Based on the sliding time-window w, carbon price forecasting results of each parameters combination are obtained. Finally, the root mean square errors (RMSE) for forecasting results of each parameter combination are calculated. RMSE is defined as $\mathrm{RMSE} = \sqrt{\frac{1}{n}\sum_{t=1}^{n}[\hat{x}(t) - x(t)]^2}$, where $x(t)$ is real value at time t, $\hat{x}(t)$ is predicted value at time t, and n is the number of training set.

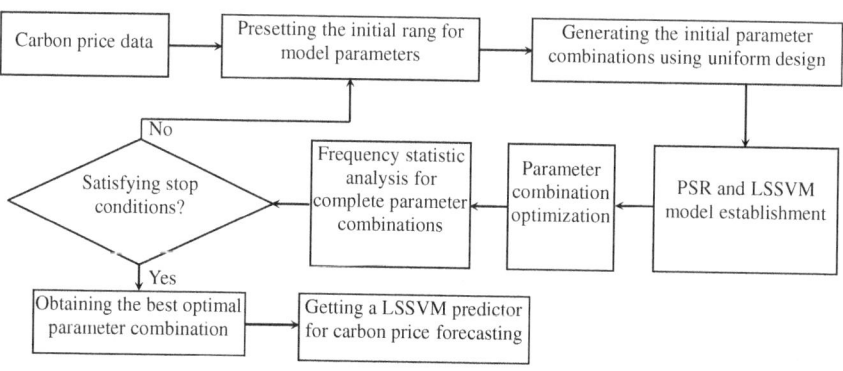

Fig. 7.1 Procedure of the proposed UD-LSSVM model

Step 4: *Parameter combination optimization.* First, a new sample set is formed by employing the obtained RMSEs as dependent variables and the corresponding parameter combinations as independent variables. Then, the sample set is utilized as a new training set of LSSVM and complete parameter combinations as the new testing set. There are merely independent variables in the testing set, while the dependent variables (RMSEs) are unknown. Finally, the new training samples are learned and the new testing set is predicted using LSSVM again to obtain the dependent variables (RMSEs) for each parameter combination in the complete parameter combinations.

Step 5: *Frequency statistical analysis for the complete parameter combinations.* The RMSEs corresponding to the complete parameter combinations are ranked, and optimized by exploring the frequencies. In frequency statistics, if the upper and lower limits of the preset parameters are not reasonable, another round of uniform design is carried out and Step 2 is performed again. Otherwise, the parameter combination corresponding to the optimal (lowest) dependent variable (RMSE) is applied as the optimal parameter combination for the LSSVM model. In this way, the optimal parameter combination for the LSSVM model is obtained.

Step 6: *Obtaining the carbon price forecasting results.* The testing set of carbon price is tested independently using the built LSSVM predictor with the obtained optimal parameters to acquire the final prediction results, so as to verify the validity of the proposed method.

7.3 Carbon Forecasting Results and Analyses

7.3.1 Data

This research uses the daily European Union Allowance (EUA) carbon futures prices with maturities of December 2015 (DEC15) and December 2016 (DEC16) in the Intercontinental Exchange (ICE) which has largest trade volume under the EU ETS as the studying samples. Considering the accessibility and continuity of the samples, carbon prices from November 29, 2011 to December 24, 2014 and November 27, 2012 to January 30, 2015 are adopted for DEC15 and DEC16, respectively. In this way, 787 and 555 sample data are obtained for DEC15 and DEC16, respectively, which are illustrated in Figs. 7.2 and 7.3 with unit of Euro/ton of carbon dioxide equivalent.

For the sake of forecast modeling, the sample data are divided into two subsets, namely, training set and testing set. The former is used to establish forecasting model while the latter is adopted to test the forecasting capacity of the built model. For DEC15, the daily carbon prices from November 29, 2011 to July 31, 2014, total 683 data, are applied as the training set and the remains from August 1, 2014 to

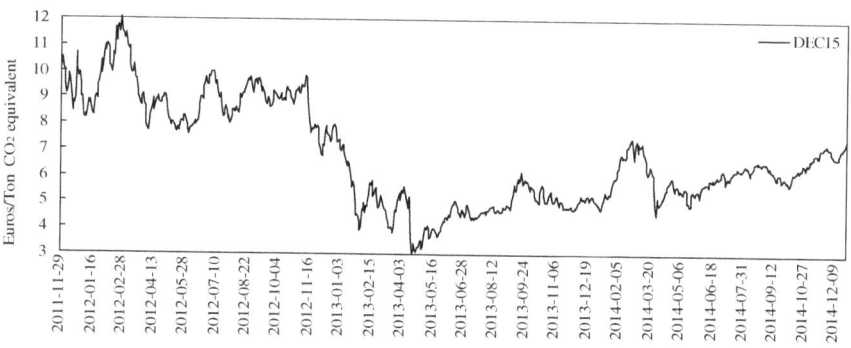

Fig. 7.2 The evolution of daily DEC15

Fig. 7.3 The evolution of daily DEC16

December 24, 2014, total 104 data, are used as the testing set. Regarding DEC16, the training set contains the daily carbon prices (455 data) from November 27, 2012 to September 9, 2014 and the testing set includes those (100 data) from September 10, 2014 to January 30, 2015. All the models are performed using a personal desktop computer with a Windows 7 system (32 bits), 4 CPU 3.00 GHZ Phenom (TM) B40 Processors, and 4.0 GB RAM.

7.3.2 Evaluation Criteria

To evaluate the forecasting capacity of the proposed models, RMSE and directional prediction statistics (D_{stat}) [13] are utilized as the evaluation criteria for level forecasting and directional prediction respectively.

Both RMSE and D_{stat} can be calculated by the forecasting results of one model. They fail to answer the question that whether model A can show a significantly higher forecasting accuracy than model B. Therefore, the Diebold–Mariano (DM) (Diebold and Mariano 1995) test and ratio test (RT) are also introduced into resolving the problem.

7.3.3 Establishment of the UD-LSSVM Model

7.3.3.1 First Round UD

The upper and lower limits of each parameter and level number are preset, as displayed in Table 7.1. A UD of five factors (parameters) and 120 levels (parameter combinations) is performed for m, τ, γ, σ, and w, as shown in Table 7.2.

7.3.3.2 Modeling of Parameter Combinations of First Round UD

Each parameter combination is trained respectively by the training sets, and tested by the testing sets of DEC15 and DEC16 through LSSVM to obtain the predicted RMSEs of the training sets, as demonstrated in Tables 7.3 and 7.4.

7.3.3.3 Optimization of Complete Parameter Combinations in First Round UD

The RMSEs of 120 samples in Tables 7.3 and 7.4 are applied as the dependent variable and m, τ, γ, σ, and w are used as the independent variables to form a sample set, which is considered as a new training set. Meanwhile, the complete combinations of five parameters including m, τ, γ, σ, and w can form a sample set, which is employed as a new testing set. The new testing set contains 28,800 (10 × 3 10 × 12 × 8) samples, among which there are merely independent variables of five parameters including m, τ, γ, σ, and w, while the dependent variable, RMSE, is unknown. By training the new training set and testing the new testing set through LSSVM, the RMSE of each parameter combination in the new testing set is obtained. In other words, 28,800 RMSEs are acquired. Those RMSEs are ranked in ascending order and the frequencies of parameter combinations corresponding to the front 2000 RMSEs are counted, as displayed in Tables 7.5 and 7.6. Here, frequency represents the times each parameter occurs.

Table 7.1 Present upper and lower limits of each parameter

Factor	m	τ	γ	σ^2	w
Upper limit	11	3	100,000	50,000	400
Lower limit	2	1	1	0.1	50
Level	10	3	10	12	8

Table 7.2 First round UD results

No.	m	τ	γ	σ^2	w	No.	m	τ	γ	σ^2	w	No.	m	τ	γ	σ^2	w
1	6	2	1	5	150	41	2	1	100,000	1	50	81	11	1	100,000	5000	350
2	3	2	10	0.1	50	42	10	2	1000	10	350	82	2	2	5000	1	300
3	6	1	10	0.5	200	43	11	3	50	1000	300	83	9	1	1000	0.5	100
4	7	2	500	10,000	100	44	4	1	500	10	300	84	5	2	1000	100	300
5	5	1	10,000	0.1	250	45	7	3	10,000	10	300	85	5	1	100,000	10,000	150
6	4	1	10	5	300	46	4	3	10	1	250	86	3	3	5000	0.1	350
7	4	3	100	5	100	47	9	2	50,000	0.1	150	87	3	3	1	10	150
8	9	3	5000	50,000	100	48	8	2	50	0.5	350	88	10	1	50,000	5	150
9	4	3	10,000	5000	150	49	9	3	100,000	1000	100	89	3	2	100	5000	400
10	5	1	100	100	100	50	11	1	10	0.1	300	90	10	1	500	1000	400
11	4	2	100,000	10	200	51	11	2	100,000	5	250	91	6	3	1000	50,000	400
12	4	2	10,000	5000	350	52	6	1	50	10	50	92	3	1	1000	10,000	250
13	9	1	50,000	50,000	400	53	10	3	100	50	50	93	6	2	50,000	50	100
14	10	3	10,000	0.5	400	54	10	2	5000	10	200	94	5	3	10	5000	250
15	5	2	50	0.5	100	55	3	2	50,000	5	400	95	2	2	10,000	500	300
16	8	3	1	100	50	56	5	3	50,000	500	150	96	8	1	10,000	10,000	200
17	2	3	500	50	150	57	11	3	50,000	50	350	97	8	2	500	50	250
18	11	2	5000	1	50	58	3	1	500	0.1	150	98	8	2	100,000	500	200
19	8	2	10	1000	350	59	6	2	5000	1000	50	99	6	3	1	500	250
20	3	3	50	500	400	60	11	1	500	1	200	100	10	2	50,000	5000	300
21	11	2	10	10,000	150	61	5	3	100	50	350	101	4	2	1	50,000	50
22	7	1	1000	10,000	250	62	2	2	50	100	200	102	6	3	100,000	50,000	300
23	2	1	50,000	0.5	250	63	5	2	1000	5	100	103	3	1	100	5000	200
24	11	1	50	5000	100	64	9	3	100	0.1	300	104	9	1	5000	5	350

(continued)

Table 7.2 (continued)

No.	m	τ	γ	σ^2	w	No.	m	τ	γ	σ^2	w	No.	m	τ	γ	σ^2	w
25	3	1	5000	1000	150	65	3	3	1000	5	50	105	2	3	500	1000	200
26	8	3	500	5000	50	66	7	1	50,000	500	50	106	10	3	10	0.5	100
27	8	3	50,000	1	300	67	10	1	100	100	150	107	10	3	5000	10,000	250
28	2	3	50,000	50,000	100	68	5	3	10,000	1	200	108	6	1	5000	500	300
29	6	1	100	1	350	69	9	2	100	50,000	150	109	10	1	1	10	250
30	7	2	50	1	150	70	8	1	1000	1	400	110	9	2	100	500	250
31	11	3	500	100	400	71	7	3	10	50,000	200	111	7	2	10,000	1000	400
32	2	2	1	10,000	300	72	6	2	100,000	0.1	400	112	7	3	1	10	400
33	4	3	5000	1000	350	73	8	1	50	50,000	300	113	5	2	500	10,000	400
34	8	1	10,000	10	100	74	4	1	10,000	50	50	114	11	3	1000	500	200
35	9	1	1	500	50	75	8	3	1000	0.1	150	115	7	3	100,000	0.5	50
36	4	2	1000	5000	50	76	9	3	1	5	350	116	4	1	100,000	100	350
37	11	2	10,000	100	100	77	10	2	1	10,000	200	117	2	1	10	1000	100
38	2	2	100	0.5	400	78	7	1	10	100	400	118	7	2	500	0.5	250
39	6	3	50	0.1	200	79	3	3	100,000	100	250	119	5	1	50	50,000	350
40	7	1	5000	50	200	80	9	2	50	50	250	120	2	1	1	50	350

Table 7.3 RMSEs of parameter combinations in first round UD for DEC15

No.	RMSE	No.	RMSE	No.	RMSE	No.	RMSE	No.	RMSE	No.	RMSE
1	0.1671	21	0.3446	41	0.5577	61	0.6857	81	0.1213	101	0.4082
2	0.2823	22	0.1202	42	0.1113	62	0.1374	82	0.1122	102	0.1185
3	0.1650	23	0.1250	43	0.1292	63	0.1470	83	0.1739	103	0.1530
4	0.1507	24	0.1912	44	0.0898	64	0.1115	84	0.4776	104	0.1494
5	0.2140	25	0.1106	45	0.1129	65	0.1371	85	0.1849	105	0.1142
6	0.1109	26	0.1545	46	0.1052	66	0.1196	86	0.1400	106	0.2986
7	0.1198	27	0.2181	47	0.4496	67	0.5179	87	0.1123	107	0.1260
8	0.1500	28	0.1132	48	0.1374	68	0.1592	88	0.2772	108	0.1090
9	0.1106	29	0.1321	49	0.1048	69	0.1305	89	0.2846	109	0.1225
10	0.1126	30	0.1824	50	0.4371	70	0.4893	90	0.1664	110	0.1276
11	0.1206	31	0.1405	51	0.1717	71	0.2238	91	0.5912	111	0.1210
12	0.1120	32	0.7812	52	0.1148	72	0.1302	92	0.2856	112	0.1546
13	0.1131	33	0.1169	53	0.1121	73	0.1267	93	0.3314	113	0.1321
14	0.1892	34	0.2176	54	0.1492	74	0.1753	94	0.1381	114	0.1201
15	0.2275	35	0.2847	55	0.1005	75	0.1121	95	0.5074	115	0.2682
16	0.2285	36	0.1279	56	0.0968	76	0.1156	96	0.1651	116	0.1078
17	0.1125	37	0.1566	57	0.1171	77	0.1523	97	0.6127	117	0.2009
18	0.2523	38	0.1170	58	0.2835	78	0.3438	98	0.1176	118	0.1842
19	0.1866	39	0.4663	59	0.1054	79	0.1224	99	0.1103	119	0.3955
20	0.1209	40	0.1106	60	0.1473	80	0.1836	100	0.1302	120	0.1464

Table 7.4 RMSEs of parameter combinations in first round UD for DEC16

No.	RMSE	No.	RMSE	No.	RMSE	No.	RMSE	No.	RMSE	No.	RMSE
1	0.1671	21	0.3446	41	0.6857	61	0.1213	81	0.1158	101	0.4082
2	0.2823	22	0.1202	42	0.1374	62	0.1122	82	0.1219	102	0.1185
3	0.1650	23	0.1250	43	0.1470	63	0.1739	83	0.2711	103	0.1530
4	0.1507	24	0.1912	44	0.1115	64	0.4776	84	0.1096	104	0.1494
5	0.2140	25	0.1106	45	0.1371	65	0.1849	85	0.1109	105	0.1142
6	0.1109	26	0.1545	46	0.1196	66	0.1400	86	0.3050	106	0.2986
7	0.1198	27	0.2181	47	0.5179	67	0.1123	87	0.1591	107	0.1260
8	0.1500	28	0.1132	48	0.1592	68	0.2772	88	0.2359	108	0.1090
9	0.1106	29	0.1321	49	0.1305	69	0.2846	89	0.1549	109	0.1225
10	0.1126	30	0.1824	50	0.4893	70	0.1664	90	0.1152	110	0.1276
11	0.1206	31	0.1405	51	0.2238	71	0.5912	91	0.1641	111	0.1210
12	0.1120	32	0.7812	52	0.1302	72	0.2856	92	0.1198	112	0.1546
13	0.1131	33	0.1169	53	0.1267	73	0.3314	93	0.1616	113	0.1321
14	0.1892	34	0.2176	54	0.1753	74	0.1381	94	0.3192	114	0.1201
15	0.2275	35	0.2847	55	0.1121	75	0.5074	95	0.1109	115	0.2682
16	0.2285	36	0.1279	56	0.1156	76	0.1651	96	0.1136	116	0.1078
17	0.1125	37	0.1566	57	0.1523	77	0.6127	97	0.1214	117	0.2009
18	0.2523	38	0.1170	58	0.3438	78	0.1176	98	0.1187	118	0.1842
19	0.1866	39	0.4663	59	0.1224	79	0.1103	99	0.3199	119	0.3955
20	0.1209	40	0.1106	60	0.1836	80	0.1302	100	0.1258	120	0.1464

Table 7.5 Least 1000 RMSEs for DEC15

m	Frequency	τ	Frequency	γ	Frequency	σ^2	Frequency	w	Frequency
2	108	1	336	1	0	0.1	83	50	0
3	108	2	336	10	0	0.5	83	100	0
4	108	3	328	50	0	1	83	150	0
5	108			100	0	5	83	200	0
6	108			500	360	10	83	250	0
7	108			1000	0	50	83	300	360
8	108			5000	0	100	83	350	0
9	100			10,000	0	500	83	400	640
10	72			50,000	280	1000	84		
11	72			100,000	360	5000	84		
						10,000	84		
						50,000	84		

Table 7.6 Least 1000 RMSEs for DEC16

m	Frequency	τ	Frequency	γ	Frequency	σ^2	Frequency	w	Frequency
2	96	1	333	1	0	0.1	84	50	0
3	105	2	344	10	0	0.5	84	100	0
4	108	3	323	50	0	1	84	150	0
5	108			100	0	5	84	200	360
6	108			500	0	10	84	250	360
7	108			1000	640	50	84	300	280
8	107			5000	0	100	84	350	0
9	105			10,000	0	500	84	400	0
10	83			50,000	0	1000	85		
11	72			100,000	360	5000	85		
						10,000	84		
						50,000	74		

7.3.3.4 Results and Discussion of First Round UD

The frequency statistic results in Tables 7.5 and 7.6 show that the frequencies of m, τ, and σ^2 are disperse, indicating that the independent variables have a slightly influence on the RMSEs, and model parameters need to be optimized further. Based on the frequency statistic results of first round UD, the ranges of several parameters are narrowed. The m of DEC15 is reduced to [2, 6] and that of DEC16 is narrowed to [3, 9]. The ranges of τ and σ^2 cannot change. The frequency statistic of γ and w is relatively concentrated. Considering the RMSEs are sensitive to the variations of γ and w, γ and w are in the ranges of [400, 800] and [240, 460] for DEC15, respectively. γ is set to 1000 and w varies in the range of [180, 315] for DEC16.

7.3.3.5 Second-Round UD Optimization

The above steps are repeated to perform the second-round optimization according to the newly parameter ranges. The results of the second-round UD optimization for DEC15 and DEC16 are illustrated in Tables 7.7, 7.8, 7.9 and 7.10 list the RMSEs of parameter combinations in the second-round UD for DEC15 and DEC16, respectively.

Tables 7.9 and 7.10 reveal that the RMSEs corresponding to parameter combinations in the second round are highly accurate. Considering the modeling efficiency and the fact that the forecasting accuracy cannot be significantly improved when the optimization continues, the optimization is stopped when high accuracy is achieved. The RMSEs corresponding to Tables 7.9 and 7.10 are ranked in ascending order and the front 20 ones are selected. The ranking results are demonstrated in Tables 7.11 and 7.12. Finally, the optimal parameter combination ($m = 5$, $\tau = 1$, $\gamma = 600$, $\sigma^2 = 100$, and $w = 340$) for DEC15 and that ($m = 4$, $\tau = 1$, $\gamma = 1000$, $\sigma^2 = 100$, and $w = 300$) for DEC16 are obtained. Using the above optimal parameters, the LSSVMs are trained for forecasting the testing sets of DEC15 and DEC16. The prediction results are illustrated in Figs. 7.4 and 7.5.

7.3.4 Comparison with PSO

To test the forecasting accuracy and efficiency of the proposed model, PSO is also employed to optimize the parameters of LSSVM (PSO-LSSVM) (Zhu and Wei 2013). In PSO, real-value encryption is applied, initial population is 100, maximum iterations is 100, acceleration factors are $c1 = c2 = 1.5$, inertia weight is $w = 1$, maximum speed is $v = 500$, and the fitness function is defined as minimum RMSE of the training set. According to the best parameters obtained in the above uniform design, carbon price data are reconstructed to be an input of LSSVM model. For DEC15, $m = 5$, $\tau = 1$ and $w = 340$; $m = 4$, $\tau = 1$, and $w = 300$ regarding DEC16. Based on those parameters, the reconstructed carbon prices are forecasted through PSO-LSSVM model. PSO is merely used to optimize the parameters including γ and σ^2 of LSSVM, and their ranges are [1, 100,000] and [0.1, 50,000], respectively. The forecasting results of various models are illustrated in Figs. 7.4 and 7.5, and the forecasting performance comparisons for both models are displayed in Tables 7.13, 7.14 and 7.15.

In terms of level forecasting, RMSE indicates that though UD-LSSVM model shows favorable level forecasting results, the results are inferior to those of PSO-LSSVM model. This is possibly because that LSSVM has a strong nonlinear modeling ability which makes it more suitable for carbon price forecasting. Moreover, the global optimization of PSO can further improve the learning and prediction capacity of LSSVM.

Table 7.7 Results of the second-round UD for DEC15

No.	m	τ	γ	σ²	w	No.	m	τ	γ	σ²	w	No.	m	τ	γ	σ²	w
1	5	2	700	500	420	41	5	3	500	5	420	81	3	1	500	0.1	320
2	2	3	600	50	260	42	2	3	800	1000	340	82	3	2	700	5000	440
3	3	3	600	1	260	43	5	2	600	10	240	83	5	1	500	1	240
4	6	2	700	50	340	44	3	3	500	1	440	84	6	2	800	5	400
5	4	2	700	10,000	320	45	6	3	600	10,000	320	85	5	3	700	10	260
6	6	3	400	5000	300	46	5	1	500	10,000	280	86	3	3	800	5000	320
7	2	1	400	10,000	260	47	6	1	400	1000	420	87	5	3	600	1000	400
8	4	2	600	500	360	48	5	2	800	10	340	88	6	2	600	10	380
9	3	3	600	5	440	49	5	2	400	50	320	89	6	1	800	0.5	300
10	4	3	800	500	420	50	4	2	600	10	380	90	6	3	500	500	320
11	4	3	500	0.5	280	51	3	3	700	1000	380	91	4	3	700	0.5	300
12	2	3	600	100	300	52	2	3	400	5000	400	92	4	3	400	1	380
13	5	2	500	5	400	53	5	3	700	50	440	93	4	1	400	50	400
14	4	1	700	5000	340	54	3	1	800	10	280	94	5	3	400	50,000	260
15	5	3	500	0.1	300	55	3	3	400	10	340	95	5	2	800	50	320
16	2	2	400	10	300	56	6	1	700	50	440	96	4	3	400	50	460
17	6	2	800	1000	280	57	2	2	500	1	440	97	2	1	700	10,000	320
18	5	3	800	1000	460	58	4	2	400	50,000	320	98	6	2	500	10	260
19	4	1	500	500	240	59	2	2	700	10	240	99	4	1	700	10,000	400
20	2	2	800	100	380	60	3	3	600	10,000	400	100	5	1	600	10,000	380
21	4	1	800	10,000	300	61	3	3	700	100	240	101	2	1	700	0.1	300
22	3	2	500	5	300	62	2	2	600	5000	420	102	5	1	600	0.5	340
23	6	2	400	5000	440	63	2	3	500	0.5	420	103	2	1	500	100	300
24	4	1	600	50,000	440	64	5	2	500	5	460	104	5	1	800	1	260

(continued)

Table 7.7 (continued)

No.	m	τ	γ	σ^2	w	No.	m	τ	γ	σ^2	w	No.	m	τ	γ	σ^2	w
25	5	3	600	100	360	65	3	2	700	1000	340	105	3	1	500	50	360
26	4	3	700	5	460	66	4	1	400	0.5	260	106	6	3	700	1	380
27	3	2	500	5	240	67	2	1	700	0.1	360	107	6	1	600	1	300
28	3	1	600	100	260	68	4	3	500	5000	360	108	2	3	800	100	280
29	5	1	800	1	380	69	4	2	700	1	240	109	6	3	800	0.5	360
30	4	1	800	50,000	420	70	6	2	400	10,000	280	110	5	1	700	500	240
31	2	2	500	500	360	71	6	1	700	0.1	440	111	3	2	400	0.5	380
32	6	3	800	5	240	72	5	2	400	0.5	460	112	2	3	400	10,000	340
33	6	3	700	0.5	280	73	3	2	800	0.1	420	113	2	2	600	100	280
34	4	2	800	50,000	260	74	6	2	600	50,000	420	114	4	2	500	100	460
35	2	1	500	50	380	75	4	3	400	1000	280	115	3	1	400	0.1	440
36	2	3	800	1000	460	76	5	1	400	500	360	116	6	1	500	100	360
37	3	2	400	5	240	77	6	1	600	0.1	460	117	6	3	500	50,000	400
38	2	1	400	500	420	78	3	1	600	50,000	280	118	3	1	800	500	400
39	6	1	400	50,000	340	79	4	2	600	0.1	340	119	3	1	600	1000	460
40	2	2	700	5000	400	80	2	1	500	50,000	320	120	3	2	800	5000	360

Table 7.8 Results of the second-round UD for DEC16

No.	m	τ	γ	σ²	w	No.	m	τ	γ	σ²	w	No.	m	τ	γ	σ²	w
1	3	2	1000	1	225	25	5	2	1000	5	240	49	9	2	1000	10,000	195
2	7	2	1000	1000	240	26	7	1	1000	5000	180	50	5	2	1000	500	270
3	6	2	1000	10	180	27	3	1	1000	50	210	51	3	2	1000	100	195
4	3	1	1000	5000	255	28	4	2	1000	50	315	52	7	2	1000	100	300
5	6	2	1000	10,000	255	29	7	1	1000	500	315	53	6	1	1000	0.1	225
6	9	1	1000	5	315	30	5	2	1000	5000	315	54	6	1	1000	1000	300
7	5	1	1000	10	300	31	9	1	1000	5	180	55	8	2	1000	1000	255
8	8	1	1000	10,000	300	32	3	2	1000	10,000	270	56	9	1	1000	5000	240
9	7	1	1000	10	255	33	5	2	1000	1	180	57	5	1	1000	10,000	210
10	6	2	1000	500	225	34	6	1	1000	100	285	58	7	2	1000	0.1	195
11	3	2	1000	1000	285	35	4	2	1000	1000	180	59	4	1	1000	10,000	285
12	9	1	1000	1000	270	36	8	2	1000	50	180	60	9	2	1000	100	210
13	8	2	1000	5000	315	37	5	2	1000	100	255	61	6	2	1000	5	210
14	5	1	1000	0.1	285	38	4	2	1000	50	240	62	3	1	1000	500	180
15	7	1	1000	50	210	39	9	2	1000	10	225	63	7	2	1000	10,000	225
16	6	1	1000	5000	195	40	8	1	1000	0.1	210	64	4	2	1000	0.1	270
17	4	1	1000	5	255	41	3	1	1000	0.1	315	65	4	1	1000	1	195
18	7	1	1000	1	270	42	3	1	1000	5	240	66	6	2	1000	1	315
19	5	1	1000	10	195	43	9	2	1000	0.1	255	67	8	1	1000	500	195
20	9	1	1000	50	285	44	8	1	1000	1	240	68	9	2	1000	500	285
21	8	1	1000	100	225	45	8	2	1000	10	270	69	4	1	1000	500	240
22	7	2	1000	5	285	46	4	1	1000	100	300	70	8	2	1000	1	300
23	3	2	1000	10	300	47	6	1	1000	50	270						
24	5	1	1000	1000	225	48	4	2	1000	5000	210						

Table 7.9 RMSEs of parameter combinations in the second-round UD for DEC15

No.	RMSE	No.	RMSE	No.	RMSE	No.	RMSE	No.	RMSE	No.	RMSE
1	0.0956	21	0.1053	41	0.1090	61	0.0916	81	0.1432	101	0.0975
2	0.0925	22	0.1002	42	0.0936	62	0.1025	82	0.1005	102	0.0878
3	0.1183	23	0.1090	43	0.1030	63	0.1032	83	0.1315	103	0.1009
4	0.0971	24	0.1518	44	0.1086	64	0.1042	84	0.1036	104	0.1362
5	0.1065	25	0.1010	45	0.1188	65	0.0934	85	0.1067	105	0.0890
6	0.1140	26	0.1035	46	0.1128	66	0.1311	86	0.0988	106	0.1266
7	0.0935	27	0.0999	47	0.0956	67	0.1326	87	0.0994	107	0.1367
8	0.0986	28	0.0887	48	0.1015	68	0.1048	88	0.1008	108	0.0920
9	0.0986	29	0.1204	49	0.0935	69	0.1355	89	0.1831	109	0.1670
10	0.0972	30	0.1427	50	0.0936	70	0.1192	90	0.1037	110	0.0888
11	0.1321	31	0.0932	51	0.0945	71	0.1989	91	0.1352	111	0.1143
12	0.0918	32	0.1039	52	0.1066	72	0.1371	92	0.1140	112	0.1191
13	0.1032	33	0.1692	53	0.1020	73	0.2101	93	0.0888	113	0.0922
14	0.0999	34	0.1456	54	0.0921	74	0.1655	94	0.1941	114	0.0936
15	0.2277	35	0.0908	55	0.0952	75	0.0963	95	0.0936	115	0.1125
16	0.0954	36	0.0956	56	0.0891	76	0.0903	96	0.0959	116	0.0887
17	0.0965	37	0.0998	57	0.0972	77	0.1974	97	0.1082	117	0.1820
18	0.0960	38	0.0941	58	0.1806	78	0.1521	98	0.1036	118	0.0896
19	0.0891	39	0.1635	59	0.0979	79	0.1888	99	0.1084	119	0.0935
20	0.0922	40	0.0998	60	0.1126	80	0.1829	100	0.2092	120	0.0980

Table 7.10 RMSEs of parameter combinations in the second-round UD for DEC16

No.	RMSE	No.	RMSE	No.	RMSE	No.	RMSE	No.	RMSE	No.	RMSE
1	0.1332	13	0.1242	25	0.1257	37	0.1084	49	0.1253	61	0.1244
2	0.1198	14	0.2162	26	0.1247	38	0.1081	50	0.1118	62	0.1141
3	0.1206	15	0.1108	27	0.1102	39	0.1426	51	0.1086	63	0.1283
4	0.1138	16	0.1191	28	0.1083	40	0.4276	52	0.1209	64	0.2728
5	0.1249	17	0.1128	29	0.1091	41	0.2485	53	0.2829	65	0.1596
6	0.1347	18	0.1633	30	0.1171	42	0.1138	54	0.1100	66	0.1935
7	0.1145	19	0.1158	31	0.1662	43	0.4362	55	0.1218	67	0.1107
8	0.1218	20	0.1103	32	0.1192	44	0.1841	56	0.1155	68	0.1268
9	0.1209	21	0.1078	33	0.2040	45	0.1415	57	0.1232	69	0.1090
10	0.1171	22	0.1483	34	0.1092	46	0.1075	58	0.4993	70	0.1956
11	0.1115	23	0.1098	35	0.1174	47	0.1107	59	0.1210		
12	0.1101	24	0.1102	36	0.1234	48	0.1165	60	0.1219		

Table 7.11 20 least RMSEs corresponding to UD parameter combinations for DEC15

m	τ	γ	σ^2	w	RMSE	Rank
5	**1**	**600**	**100**	**340**	**0.0878**	**1**
3	1	600	100	260	0.0887	2
6	1	500	100	360	0.0887	3
5	1	700	500	240	0.0888	4
4	1	400	50	400	0.0888	5
3	1	500	50	360	0.0890	6
6	1	700	50	440	0.0891	7
4	1	500	500	240	0.0891	8
3	1	800	500	400	0.0896	9
5	1	400	500	360	0.0903	10
2	1	500	50	380	0.0908	11
3	3	700	100	240	0.0916	12
2	3	600	100	300	0.0918	13
2	3	800	100	280	0.0920	14
3	1	800	10	280	0.0921	15
2	2	800	100	380	0.0922	16
2	2	600	100	280	0.0922	17
2	3	600	50	260	0.0925	18
2	2	500	500	360	0.0932	19
3	2	700	1000	340	0.0934	20

Table 7.12 20 least RMSEs corresponding to UD parameter combinations for DEC16

m	τ	γ	σ^2	w	RMSE	Rank
4	**1**	**1000**	**100**	**300**	**0.1075**	**1**
8	1	1000	100	225	0.1078	2
4	2	1000	50	240	0.1081	3
4	2	1000	50	315	0.1083	4
5	2	1000	100	255	0.1084	5
3	2	1000	100	195	0.1086	6
4	1	1000	500	240	0.1090	7
7	1	1000	500	315	0.1091	8
6	1	1000	100	285	0.1092	9
3	2	1000	10	300	0.1098	10
6	1	1000	1000	300	0.1100	11
9	1	1000	1000	270	0.1101	12
3	1	1000	50	210	0.1102	13
5	1	1000	1000	225	0.1102	14
9	1	1000	50	285	0.1103	15
6	1	1000	50	270	0.1107	16
8	1	1000	500	195	0.1107	17
7	1	1000	50	210	0.1108	18
3	2	1000	1000	285	0.1115	19
5	2	1000	500	270	0.1118	20

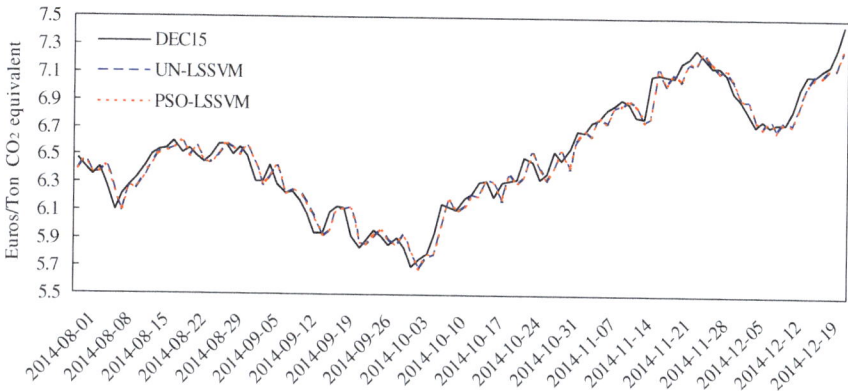

Fig. 7.4 Forecasting results for DEC15

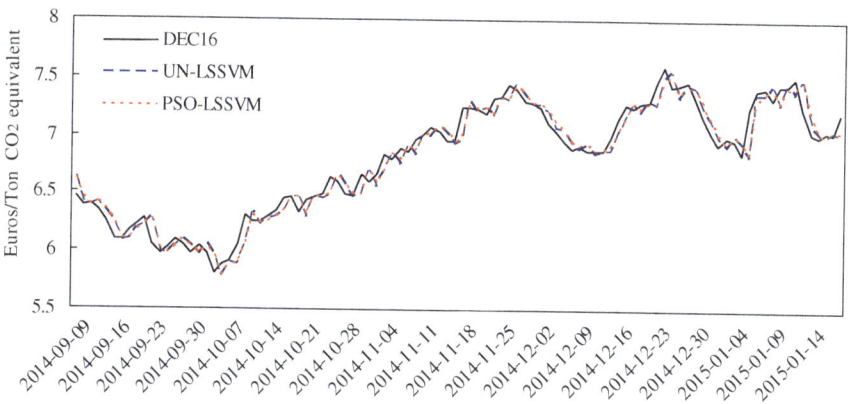

Fig. 7.5 Forecasting results for DEC16

Table 7.13 Comparisons between UD-LSSVM and PSO-LSSVM

Model	DEC15			DEC16		
	RMSE	D_{stat}	Time (s)	RMSE	D_{stat}	Time (s)
UD-LSSVM	0.0875	67.31	89.31	0.1075	63	68.29
PSO-LSSVM	0.0872	68.27	100568.27	0.1066	67	97206.02

Table 7.14 DM test for UD-LSSVM and PSO-LSSVM

Test model	Reference model: PSO-LSSVM	
	DEC15	DEC16
UD-LSSVM	−0.5354(0.5924)*	−0.5866(0.5575)*

*z_{DM}(*p-value*)

Table 7.15 RT test for UD-LSSVM and PSO-LSSVM

Model	PSO-LSSVM	
	DEC15	DEC16
UD-LSSVM	−0.1482(0.8822)*	−0.5935(0.5528)*

*z_{RT} (*p-value*)

From the aspect of directional prediction, similar conclusions to RMSE are drawn according to D_{stat} results. That is, though UD-LSSVM model has preferable directional prediction results, the results are inferior to those of PSO-LSSVM model.

Considering the modeling efficiency, the computation time of both the models reveals that UD-LSSVM model displays significant advantages and much higher efficiency in the optimization of parameter combination comparing with PSO-LSSVM model. This is mainly because, besides ensuring a high optimization accuracy, uniform design can select point set of low deviation and uniform distribution in the experimental range to set experimental points in the parameter optimization. In this way, the experimental times are dramatically decreased, so the proposed method is more suitable for optimization of parameter combination of a complex model with input of multiple parameters, and can efficiently transform the large sample search for optimization of multiple parameters into a small sample search.

Regarding DM test, the test results indicate that though the level forecasting results of UD-LSSVM model are slightly inferior to those of PSO-LSSVM model, the accuracies of level forecasting results of those two models have no difference at a significance level of 5%.

Considering RT test, the test results show that though the directional prediction results of UD-LSSVM model are worse than those of PSO-LSSVM model, the accuracies of directional forecasting results of those two models display no difference with a significance of 5%.

In summary, the results of empirical analysis for DEC15 and DEC16 indicate that: first, UD-LSSVM can achieve favorable forecasting accuracies in both level and directional forecasts. Second, UD-LSSVM can remarkably improve the modeling efficiency of carbon price forecasting. Third, at the same time ensuring a high prediction accuracy, UD-LSSVM can significantly improve the modeling efficiency for forecasting, and thereby can provide a competitive solution for carbon price forecasting.

7.4 Conclusion

According to Standard & Poor's, nearly 40% of business activities nowadays are impacted by compliance liabilities under some sort of environmental, energy, or sustainability regulatory scheme. At the heart of operations management, scheduling and planning lie the accurate forecast of the said liability. Against the background of carbon markets, carbon price constitutes the key variable of interest for regulated companies. To meet their expectations, we develop a model offering high performance in terms of carbon price forecasting.

This chapter proposes a uniform design-based least squares support vector machine (UD-LSSVM) for carbon price forecasting by simultaneously optimizing phase space reconstruction and model parameters of LSSVM using uniform design. The proposed method is verified by forecasting two carbon futures prices with different maturities under the EU ETS. The results reveal that: first, by simultaneously optimizing the phase space reconstruction and model parameters of LSSVM utilizing uniform design, model parameters are selected more reasonably without requiring for any prior knowledge, and avoiding the defects of traditional separate optimization and alternative optimization. Second, uniform design can handle the combined optimization of multiple factors and multiple levels by transforming the large sample search of simultaneously optimization into a small sample search. Besides the insurance of a high prediction accuracy, uniform design can largely reduce the computation complexity of model parameter selection, and accelerate the training speed for LSSVM predictor. Therefore, the proposed method is a universal, rapid, effective, and generalized modeling approach for carbon price forecasting, which can be used as a competitive alternative scheme for future carbon price forecasting in China. However, how to construct the optimal prediction model according to characteristics of carbon price data so as to further improve the forecasting accuracy is one of the future research tasks.

References

Albano AM, Muench J, Schwartz C et al (2002) Correlation-dimension and autocorrelation fluctuations in epileptic seizure dynamics. Phys Rev 65(3):1921–1926

Benz E, Truck S (2009) Modeling the price dynamics of CO_2 emission allowances. Energy Econ 31(1):4–15

Byun SJ, Cho H (2013) Forecasting carbon futures volatility using GARCH models with energy volatilities. Energy Econ 40:207–221

Cao L (1997) Practical method for determining the minimum embedding dimension of a scalar time series. Physica D 110:43–50

Chevallier J (2010) Volatility forecasting of carbon prices using factor models. Econ Bull 30 (2):1642–1660

Chevallier J (2011) Nonparametric modeling of carbon prices. Energy Econ 33(6):1267–1282

Chevallier J, Sevi B (2011) On the realized volatility of the ECX emissions 2008 futures contract: distribution, dynamics and forecasting. Ann Finance 7:1–29

Conrad C, Rittler D, Rotfub W (2012) Modeling and explaining the dynamics of European Union Allowance prices at high-frequency. Econ Bull 34(1):316–326

Diebold FX, Mariano RS (1995) Comparing predictive accuracy. J Bus Econ Stat 13(3):253–263

Fan XH, Li S, Tian LX (2015) Chaotic characteristic identification for carbon price and an multi-layer perception network prediction model. Expert Syst Appl 42:3945–3952

Fang KT (1994) Uniform design & uniform design table. Science Press, Beijing

Fraser AM (1989) Information and entropy in strange attractors. IEEE Trans Inf Theory 35: 245–262

Kanen JLM (2006) Carbon trading and pricing. Environmental Finance Publications, London

Kim HS, Eykholt R, Salas JD (1999) Nonlinear dynamics, delay times and embedding windows. Physica D 127:48–60

Koop G, Tole L (2013) Forecasting the European carbon market. J Roy Stat Soc Ser A, Roy Stat Soc 176(3):723–741

Liu XY, Shao C, Ma HF, Liu RX (2011) Optimal earth pressure balance control for shield tunneling based on LS-SVM and PSO. Autom Constr 20:321–327

Maguire LP, Roche B, Mcginnity TM (1998) Predicting a chaotic time series using a fuzzy neural network. Inf Sci 112:125–136

Paolella MS, Taschini L (2008) An econometric analysis of emission allowance prices. J Bank Finance 32:2022–2032

Reilly JM, Paltsev S (2005) An analysis of the European emission trading scheme. Report No. 127, MIT Joint Program on the Science and Policy of Global Change

Silva DA, Silva JP, Neto ARR (2015) Novel approaches using evolutionary computation for sparse least square support vector machines. Neurocomputing 168:908–916

Zhang YJ, Wei YM (2010) An overview of current research on EU ETS: Evidence from its operating mechanism and economic effect. Appl Energy 87(6):1804–1814

Zhang W, Niu P, Li G, Li P (2013) Forecasting of turbine heat rate with online least squares support vector machine based on gravitational search algorithm. Knowl-Based Syst 39:34–44

Zhou LG, Lai KK, Yu L (2009) Credit scoring using support vector machines with direct search for parameters selection. Soft Comput 13:149–155

Zhu BZ (2012) A novel multiscale ensemble carbon price prediction model integrating empirical mode decomposition, genetic algorithm and artificial neural network. Energies 5:355–370

Zhu BZ, Wei YM (2013) Carbon price prediction with a hybrid ARIMA and least squares support vector machines methodology. Omega 41:517–524

Chapter 8
Forecasting Carbon Price with Empirical Mode Decomposition and Least Squares Support Vector Regression

Abstract This chapter contains another hybrid model of carbon price forecasting that combines empirical mode decomposition and least squares support vector regression. This multiscale prediction methodology yields accurate forecasts of the carbon futures contracts, superior to ARIMA time series models.

8.1 Introduction

The EU ETS has been the biggest carbon trading market so far, which can provide an important emission reduction tool for the populace as well as a major investment choice for investors. In light of this, it is clear that investment risks and costs can be reduced by improving the accuracy of carbon price forecasting.

In recent years, more and more scholars have paid attention to carbon price forecasting. The approaches so far applied are divided into two groups by and large: econometric approaches and artificial intelligence (AI) techniques. Econometric approaches include multiple linear regression (Guðbrandsdóttir and Haraldsson 2011), GARCH (Paolella and Taschini 2008), MS–AR–GARCH (Benz and Truck 2009) FIAPGARCH (Conrad et al. 2012), HAR–RV (Chevallier and Sevi 2011), and nonparametric models (Chevallier 2011). The AI techniques used include artificial neural networks (ANNs) and least squares support vector regression (LSSVR) models (Zhu and Wei 2013). Although these existing methods can obtain good results when applied to stationary time series forecasting, carbon prices are highly non-stationary and nonlinear (Feng et al. 2011). Thus, such methods are not suitable for carbon price forecasting.

Empirical mode decomposition (EMD), developed by Huang and his coworkers in 1998, is an effective approach for analyzing and predicting nonlinear and non-stationary time series (Huang et al. 1998, 1999). EMD can disassemble any carbon price into a batch of independent, semblable characteristic and high regular intrinsic mode functions (IMFs) and a monotonic remainder. When used as an input

Special thanks to Dong Han and Yi-Ming Wei for helping writing on Chap. 8.

for LSSVR, an IMF can improve learning efficiency and forecasting accuracy by providing better understanding and feature-capturing (Zhu 2012). Thereby, carbon price forecasting precision is enhanced using EMD. In recent years, some EMD-based ANN and LSSVR models have been applied in several studies involving time series forecasting (Yu et al. 2008; Tang et al. 2012; Chen et al. 2012; Lin et al. 2012; Wei and Chen 2012; An et al. 2013) and produced good results. Carbon price forecasting has also been addressed (Zhu 2012). However, traditional back-propagation ANNs have been mostly used as the predictors in existing studies and may this lead to overfitting problems. However, LSSVR, which is built on basis of structural risk minimization (SRM), can solve the overfitting problem (Suykenns and Vandewalle 1999). Hybrid EMD and LSSVR models have rarely been employed in predicting carbon price. Thus, this chapter is performed to address this situation.

The purpose of this chapter is to create a novel multiscale prediction model hybridizing EMD, PSO, and LSSVR to forecast carbon price and to compare it with several other forecasting methods. The main contributions of this study are twofold. First, a novel multiscale prediction model hybridizing EMD and LSSVR is proposed for predicting carbon price. To begin with, this model uses EMD for disassembling carbon price into a batch of high regular IMFs and one monotonic remainder. Then, each component is independently predicted using a LSSVR predictor trained using particle swarm optimization (PSO–LSSVR). Finally, the forecasting values of all the IMFs and remainder are aggregated into the eventual predictive values of original carbon price. Second, the individual ARIMA and LSSVR models, the hybrid ARIMA+LSSVR model, and two multiscale forecasting models (an EMD-based ARIMA model and the proposed EMD-based LSSVR model) are compared with each other using some used widely evaluation criteria such as level forecasting, directional prediction, and the Diebold–Mariano (DM) test. The outcomes reveal that the presented EMD-based LSSVR model can triumph over the other popular forecasting methods.

8.2 Methodology

8.2.1 Hybridizing EMD and LSSVR for Carbon Price Prediction

In terms of carbon price $x_t, (t = 1, 2, \ldots, T)$, a h-step forecasting in advance, i.e., x_{t+h}, can be expressed in the form (8.1) as

$$\hat{x}_{t+h} = f(x_t, x_{t-1}, \ldots, x_{t-m+1}), \qquad (8.1)$$

where x_t is the real value, $\hat{x}_t$ is the forecasted value, and m is the lag order (Fig. 8.1).

As shown in Fig. 8.2, we propose a new multiscale prediction model hybridizing EMD, PSO, and LSSVR for carbon price forecasting, which is usually comprised of the subsequent three key steps:

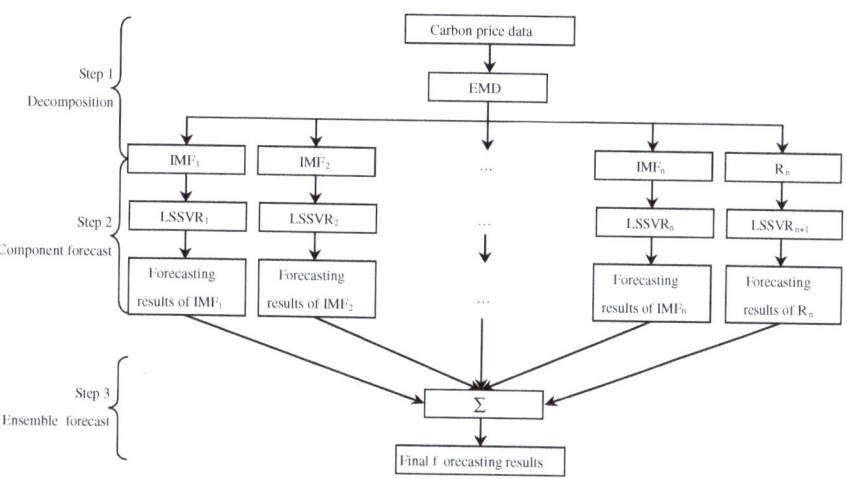

Fig. 8.1 The framework for the proposed multiscale prediction methodology

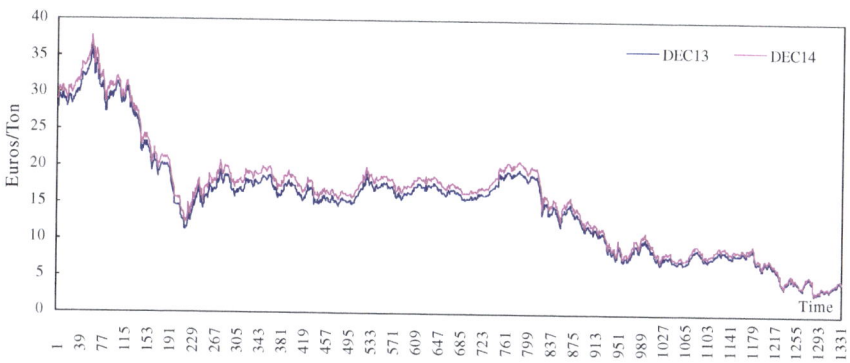

Fig. 8.2 The DEC13 and DEC14 data from April 8, 2008 to June 21, 2013

Step 1: Carbon price is disassembled into a batch of high regular IMFs and one monotonic residue via EMD.

Step 2: PSO–LSSVR is respectively employed in forecasting the IMFs and residue. Using these, we can obtain their individual forecasting results.

Step 3: The forecasting values of all the IMFs and residue are summarized into the final predictive values of original carbon price.

In short, the proposed multiscale prediction model hybridizing EMD and LSSVR is in essence an EMD (Decomposition)–LSSVR (Single forecasting)–ADD (Ensemble) model, which is a utilization of "decomposition and ensemble" tactics. In the next section, two carbon future prices are used for checking the validity of the proposed multiscale prediction approach.

8.3   Experimental Analysis

8.3.1   Carbon Prices

Two carbon future prices matured in December 2013 and 2014 referred to as DEC13 and DEC14 here, respectively, were selected as empirical samples in this chapter, as shown in Fig. 8.2 with a unit of Euros/ton. The samples, involving a total of 1335 observations and in units of Euros/ton, were gathered from April 8, 2008 to June 21, 2013 excluding public holidays. For convenience, in LSSVR modeling of the DEC13 and DEC14 data, the daily data involving 1000 data points collected from April 8, 2008 to February 29, 2012, excluding public holidays, were treated as the in-sample training samples. The remains were employed as the out-of-sample test samples in order to check the model's predictive performance. It is evident that these two carbon prices changes tend to be nonstationary and non-linear—this is due to their means changing over time. Therefore, many difficulties will definitely be encountered in accurately forecasting carbon prices.

8.3.2   Evaluation Criteria

Forecasting performance is measured by evaluating the level forecasting and directional prediction using two main criteria. Level forecasting is measured via the root mean squared error (RMSE). Directional prediction is measured through the directional prediction statistic (D_{stat}), where x_t is the real value, $\hat{x}_t$ is the predictive value, and n is the amount of test samples.

The DM test is further used to statistically contrast the predicted performances of various predictive models. In this chapter, mean square prediction error (MSPE) was chosen as the loss function.

For the sake of comparing the predictive performance of EMD–LSSVR–ADD with the popular predicted methods, this research uses individual ARIMA and LSSVR models, a hybrid ARIMA+LSSVR model, and a variant of the EMD–ARIMA–ADD model, as the benchmark models. In the hybrid ARIMA+LSSVR model, the linear part of carbon price is modeled and predicted via the ARIMA model. In addition, LSSVR is used for modeling the residual which only contains the nonlinear part of carbon price. Thus, the eventual predictive value of the original carbon price is obtained from the sum of the predictive values of both the linear and nonlinear parts. In the variant of the EMD–ARIMA–ADD model, ARIMA is independently employed in modeling the IMF components and the residue extracted using EMD. Therefore, the eventual predictive value of the original carbon price is achieved from the total of the predicted values of all the IMFs plus the residue.

8.3.3 Predicted Results

Forecasting tests were carried out according to the steps outlined in the previous section. To begin with, we set up in advance the thresholds and tolerance using $[\theta_1, \theta_2, \alpha] = [0.05, 0.5, 0.05]$. Then, Figs. 8.3 and 8.4 illustrated the decomposition results using EMD. Clearly, DEC13 and DEC14 are, respectively, disassembled into eight IMFs, seven IMFs, and one residue. At the same time, all the IMFs and residues have similar features and strong regularity.

This chapter performed one-step-ahead forecasting for DEC13 and DEC14. The EViews developed by Quantitative Micro Software, USA is used for ARIMA modeling. The optimum model is identified using the Akaike information criterion (AIC). By trial and error, both the best models derived from DEC13 and DEC14 are ARIMA (3,1,2) models. Tables 8.1 and 8.2 list the estimated results. Moreover, as previously mentioned, ARIMA is also used to model each IMF and residue decomposed by EMD. The predictive values of all IMFs and residue are then composed into the predicted values of EMD–ARIMA–ADD model.

The LSSVR model was built using the LSSVR lab produced by Suykens and his colleagues on the platform of MATLAB 2013a. The input of each LSSVR model was determined using a partial autocorrelation function method (Zhu 2012). The optimal parameters were searched for using 100 particles and 50 generations. The factors assumed were $C \in [1, 500]$, $\sigma \in (0, 50]$, $c_1 = c_2 = 2$, $w_{max} = 0.9$, $w_{min} = 0.1$, $p_{max} = 0.5$, and $v_{max} = 50$. Finally, the optimal parameters were obtained: $C = 200.1568$ and $\sigma = 39.3093$ for DEC13 and $C = 300.6328$ and $\sigma = 44.8074$ for DEC14. These were used to build LSSVR models. Moreover, as mentioned above, LSSVR is also used to model each IMF and residue decomposed using EMD, and the predictive values of all IMFs and residue are composed into the forecasting values for EMD–LSSVR–ADD model.

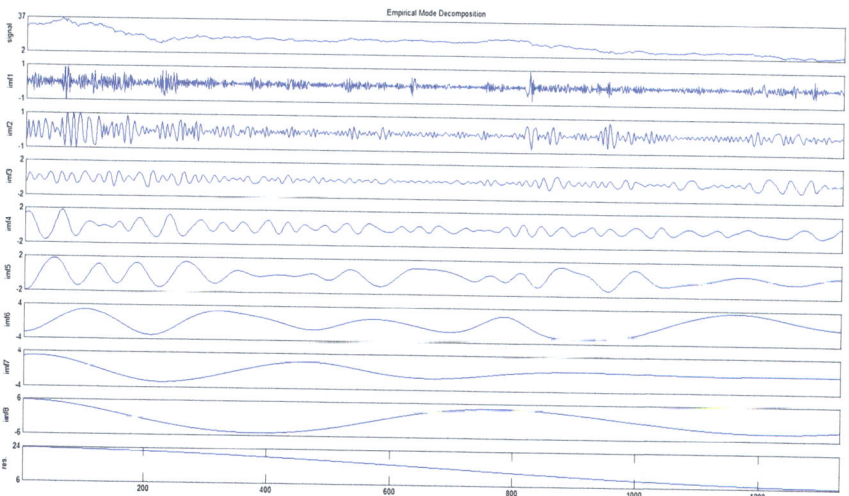

Fig. 8.3 DEC13 decomposed results using EMD

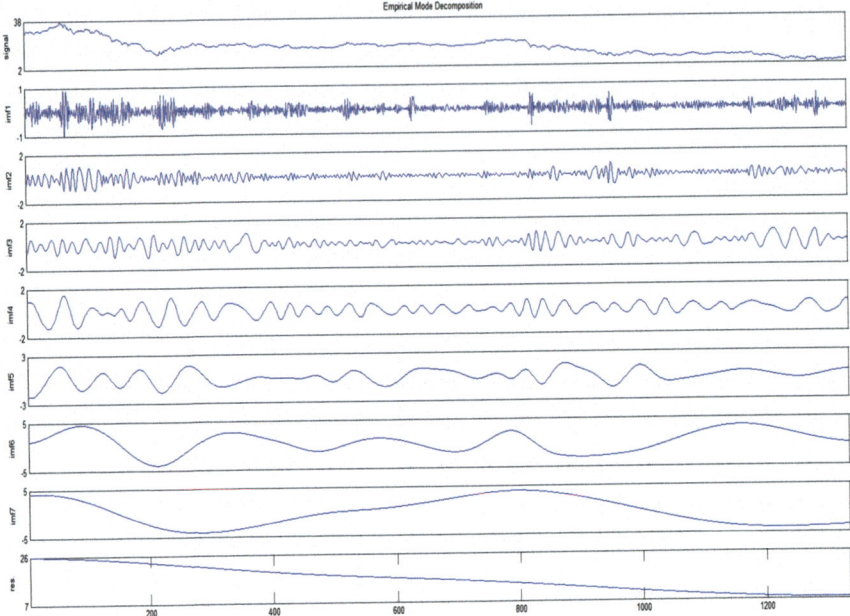

Fig. 8.4 DEC14 decomposed results using EMD

Table 8.1 The estimated ARIMA model for DEC13

Variable	Coefficient	Std. error	t-statistic	Prob.
C	−0.019631	0.014084	−1.393838	0.1637
AR(1)	0.662517	0.055186	12.00510	0.0000
AR(2)	−0.958197	0.043999	−21.77748	0.0000
AR(3)	0.140847	0.032366	4.351635	0.0000
MA(1)	−0.569881	0.046860	−12.16139	0.0000
MA(2)	0.872599	0.046366	18.81993	0.0000
R-squared	0.035676	Mean-dependent var		−0.019478
Adjusted R-squared	0.030806	S.D.-dependent var		0.400317
S.E. of regression	0.394102	Akaike info criterion		0.981594
Sum squared resid	153.7635	Schwarz criterion		1.011134
Log likelihood	−482.8337	Hannan–Quinn criter		0.992824
F-statistic	7.325172	Durbin–Watson stat		1.989720
Prob (F-statistic)	0.000001			

The hybrid model was built as discussed above. First, the final prediction for the original carbon price was obtained using ARIMA to forecast the linear element of carbon price, and employing LSSVR in predicting the nonlinear element of carbon price. Consequently, two individual models (LSSVR and ARIMA), a hybrid ARIMA +LSSVR model, and two multiscale forecasting models (EMD–ARIMA–ADD and

Table 8.2 The estimated ARIMA model for DEC14

Variable	Coefficient	Std. error	t-statistic	Prob.
C	−0.019916	0.014172	−1.405249	0.1603
AR(1)	0.651735	0.050916	12.80027	0.0000
AR(2)	−0.975131	0.037915	−25.71906	0.0000
AR(3)	0.131702	0.032397	4.065254	0.0001
MA(1)	−0.560882	0.041156	−13.62806	0.0000
MA(2)	0.893940	0.039845	22.43516	0.0000
R-squared	0.034872	Mean-dependent var		−0.019729
Adjusted R-squared	0.029997	S.D.-dependent var		0.406049
S.E. of regression	0.399912	Akaike info criterion		1.010861
Sum squared resid	158.3303	Schwarz criterion		1.040402
Log likelihood	−497.4088	Hannan–Quinn criter		1.022091
F-statistic	7.154106	Durbin–Watson stat		1.991415
Prob (F-statistic)	0.000001			

EMD–LSSVR–ADD) were applied to forecast carbon prices in this chapter. The results of RMSE and D_{stat} for the different models are shown in Figs. 8.5, 8.6, 8.7 and 8.8. The DM test results are listed in Tables 8.3 and 8.4. It is obvious that EMD–LSSVR–ADD is consistently and significantly superior to all of the other popular

Table 8.3 DM test results for DEC13

Test model	Reference model			
	EMD–ARIMA–ADD	Hybrid	LSSVR	ARIMA
EMD–LSSVR–ADD	−1.2905 (0.0984)	−4.3545 (6.7E-06)	−5.0489 (2.27E-07)	−5.8023 (3.3E-09)
EMD–ARIMA–ADD		−4.2252 (1.2E-05)	−4.8909 (5.0E-07)	−5.6497 (8E-09)
Hybrid			−2.1976 (0.0140)	−1.6792 (0.0466)
LSSVR				−1.2956 (0.0976)

Table 8.4 DM test results for DEC14

Test model	Reference model			
	EMD–ARIMA–ADD	Hybrid	LSSVR	ARIMA
EMD–LSSVR–ADD	−3.7755 (8E-05)	−7.618 (1.3E-14)	−5.8409 (2.6E-09)	−6.4649 (5.1E-11)
EMD–ARIMA–ADD		−7.5187 (2.8E-14)	−5.394 (3.4E-08)	−5.8201 (2.9E-09)
Hybrid			−1.7641 (0.0389)	−1.4462 (0.0741)
LSSVR				−1.3161 (0.0941)

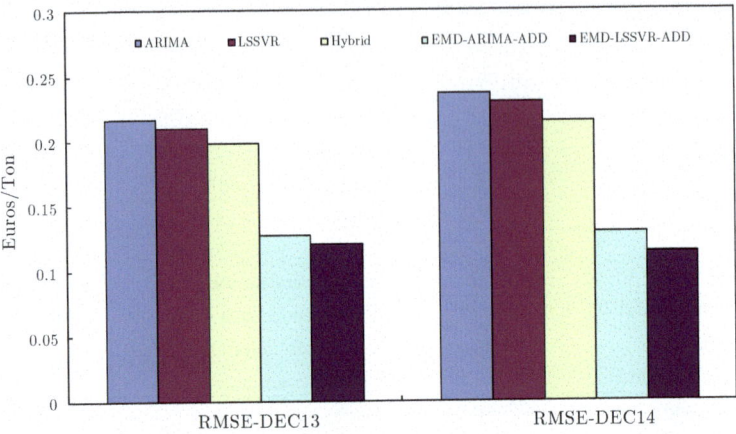

Fig. 8.5 RMSE of different forecasting models

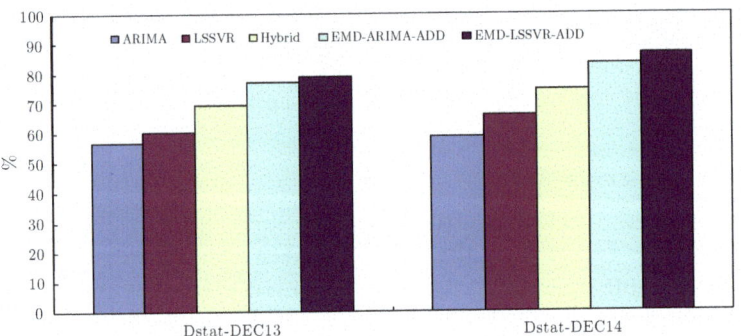

Fig. 8.6 D_{stat} of different forecasting models

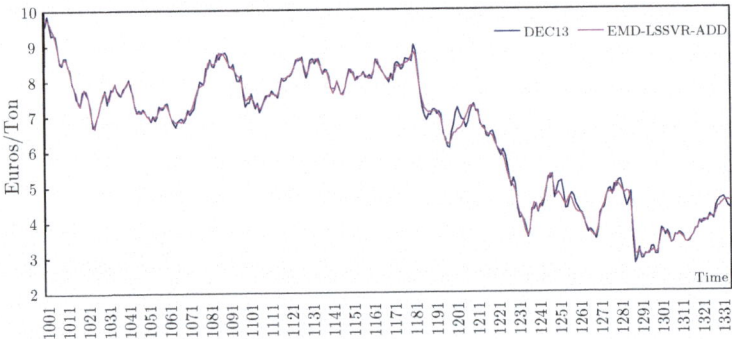

Fig. 8.7 Out-of-sample forecasting results for DEC13

Fig. 8.8 Out-of-sample forecasting results for DEC14

forecasting approaches. In all the models used, EMD–LSSVR–ADD achieves the highest measurement accuracy (as measured by RMSE) and the highest direction decision hit rate (as measured by D_{stat}). Furthermore, the DM results for the proposed model show that all the values of the DM statistics are less than -1.2905 and all the values of p are lower than 10%, which indicates that EMD–LSSVR–ADD has the best short-term carbon price forecasting behavior at a 90% significance level.

According to the RMSE values, the presented EMD–LSSVR–ADD model is the best, followed by the EMD–ARIMA–ADD, ARIMA+LSSVR, LSSVR, and ARIMA models. At the same time, ARIMA is inferior to LSSVR. The feasible justification is that ARIMA is in essential a kind of linear method, while LSSVR is a class of nonlinear model. Thus, the linear model may be unsuitable to predict carbon prices which are nonstationary and nonlinear. Moreover, PSO has enhanced the forecasting ability of LSSVR because of its global optimization capacity. Both the individual ARIMA and LSSVR models are worse than the hybrid model which may be attributed to the influence of the hybrid strategy on forecasting performance. Besides this, two multiscale forecasting models are superior to single and hybrid models due to the effects of EMD decomposition on prediction performance.

The D_{stat} results also show that EMD–LSSVR–ADD is superior to the other forecasting models. This may be attributed to the superiority of "decomposition and ensemble" tactics which has a big influence on the predictive performance. Meantime, the hybrid model also exceeds the single ARIMA and LSSVR models. One reason for this could be that a hybrid strategy can make full use of their combined advantages and thus enhances the forecasting performance. LSSVR also behaves better than ARIMA, but this is mostly because ARIMA, as a linear model, fails to capture the complex intrinsic characteristics of the highly nonlinear carbon prices.

Five findings are derived from the DM test results. First, when the test model is EMD–LSSVR–ADD, all the values of p are less than 10%. This indicates that EMD–LSSVR–ADD statistically performs better than all the other models used with a significance level of 90%. Second, the multiscale models are superior to single and hybrid models at a 99% significance level, indicating the effectiveness of EMD. Third, the proposed EMD-based LSSVR model yields better results than EMD–ARIMA–ADD at a significance level of 90%. Fourth, the AI model is significantly superior to the econometric model. Finally, it is proved that traditional forecasting approaches, i.e., without EMD preprocessing, are not capable of accurately forecasting nonstationary and nonlinear carbon prices.

To sum up, we can draw a few conclusions from the experimental results: (1) According to the evaluation criteria of level prediction, directional forecasting, and DM test, compared with ARIMA, LSSVR, ARIMA+LSSVR, and EMD–ARIMA–ADD models, EMD–LSSVR–ADD model shows superior forecasting performance. (2) EMD–LSSVR–ADD and EMD–ARIMA–ADD achieve more precise prediction results than ARIMA, LSSVR, and ARIMA+LSSVR models, which implies that "decomposition and ensemble" tactics can significantly enhance predictive capability. (3) Nonlinear approaches are more appropriate to forecast carbon price than linear models. Therefore, it is concluded that the designed multiscale predictive model hybridizing EMD and LSSVR is a very capable method with regards carbon price forecasting.

8.4 Conclusion

In this research, we present a span-new multiscale prediction model hybridizing EMD, PSO, and LSSVR to forecast carbon price, which is a promising forecasting approach to nonstationary and nonlinear carbon price. This model use EMD to disassemble carbon price into a batch of more stationary and more regular components. Thereby, the LSSVR model of each component can be easily built. The final forecasting values of carbon price are obtained from summarizing the forecasting results of all the components. Two carbon future prices from the ECX market have been evaluated. Finally, the proposed EMD-based LSSVR model is compared with individual ARIMA and LSSVR models, hybrid ARIMA+LSSVR model, and EMD–ARIMA–ADD, based on RMSE, D_{stat}, and DM test results.

The results proclaim that the proposed multiscale prediction model yields the lowest RMSE and highest D_{stat} value. Moreover, the DM test results reveal that the presented EMD–LSSVR–ADD model significantly exceeds the individual ARIMA and LSSVR models, and the hybrid ARIMA+LSSVR and EMD–ARIMA–ADD models. Therefore, the proposed multiscale prediction model can be used as a promising solution for nonlinear and nonstationary carbon price prediction.

References

An N, Zhao WG, Wang JZ, Shang D, Zhao ED (2013) Using multi-output feedforward neural network with empirical mode decomposition based signal filtering for electricity demand forecasting. Energy 49(1):279–288

Benz E, Truck S (2009) Modeling the price dynamics of CO_2 emission allowances. Energy Econ 31(1):4–15

Chen CF, Lai MC, Yeh CC (2012) Forecasting tourism demand based on empirical mode decomposition and neural network. Knowl Based Syst 26:281–287

Chevallier J (2011) Nonparametric modeling of carbon prices. Energy Econ 33(6):1267–1282

Chevallier J, Sevi B (2011) On the realized volatility of the ECX emissions 2008 futures contract: distribution, dynamics and forecasting. Ann Finance 7:1–29

Conrad C, Rittler D, Rotfub W (2012) Modeling and explaining the dynamics of European union allowance prices at the high-frequency. Energy Econ 34(1):316–326

Feng ZH, Zou LL, Wei YM (2011) Carbon price volatility: evidence from EU ETS. Appl Energy 88:590–598

Guðbrandsdóttir HN, Haraldsson HÓ (2011) Predicting the price of EU ETS carbon credits. Syst Eng Proc 1:481–489

Huang NE, Shen Z, Long SR (1998) The empirical mode decomposition and the Hilbert spectrum for nonlinear and nonstationary time series analysis. Proc R Soc Lond A454:903–995

Huang NE, Shen Z, Long SR (1999) A new view of nonlinear water waves: the Hilbert spectrum. Annu Rev Fluid Mech 31:417–457

Lin CS, Chiu SH, Lin TY (2012) Empirical mode decomposition–based least squares support vector regression for foreign exchange rate forecasting. Econ Model 29(6):2583–2590

Paolella MS, Taschini L (2008) An econometric analysis of emission allowance prices. J Bank Finance 32:2022–2032

Suykenns JAK, Vandewalle J (1999) Least squares support vector machine. Neural Process Lett 9(3):293–300

Tang L, Yu LA, Wang S, Li JP, Wang SY (2012) A novel hybrid ensemble learning paradigm for nuclear energy consumption forecasting. Appl Energy 93:432–443

Wei Y, Chen MC (2012) Forecasting the short-term metro passenger flow with empirical mode decomposition and neural networks. Transp Res Part C Emerg Technol 21(1):148–162

Yu LA, Wang SY, Lai KK (2008) Forecasting crude oil price with an EMD-based neural network ensemble learning paradigm. Energy Econ 30(5):2623–2635

Zhu BZ (2012) A novel multiscale ensemble carbon price prediction model integrating empirical mode decomposition, genetic algorithm and artificial neural network. Energies 5:355–370

Zhu BZ, Wei YM (2013) Carbon price prediction with a hybrid ARIMA and least squares support vector machines methodology. Omega 41:517–524

Chapter 9
An Adaptive Multiscale Ensemble Learning Paradigm for Carbon Price Forecasting

Abstract This final chapter is devoted to an adaptive model of carbon price forecasting that makes use of artificial neural networks. Considering either ensemble empirical mode decomposition, the least squares support vector machine, or the particle swarm optimization variant, the competing models are given an extra dimension by incorporating a learning paradigm.

9.1 Introduction

The accurate prediction on carbon price, on one hand, can contribute to a deep understanding on the fluctuation mechanism of carbon price so as to ensure the economic and energy security. On the other hand, it is beneficial for production operations and investment decisions so as to help avoid risks posed on carbon price to achieve the maximum profit.

The violent and frequent fluctuation of carbon price gives rise to its complex characteristics of high nonstationarity, nonlinearity, and chaos, which can pose a great challenge for carbon price prediction. Carbon price forecasting is not only a research hot pot but also a hard issue in the academic world. Although the research methods for forecasting carbon price tend to be diversified, they can be roughly divided into two categories: one is the statistical prediction method represented by the econometric models. The existing results show that the statistical prediction method is capable of predicting the carbon price with a high accuracy. However, carbon price fluctuates in a highly nonlinear and nonstationary way, while the traditional statistical and econometric models are established on the assumption that data are stationary and linear. As a result, this method fails to effectively deal with the nonlinear patterns concealed in the nonstationary carbon price, which can lead to the unsatisfied prediction results.

Special thanks to Xuetao Shi, Ping Wang, Dong Han and Ying-Ming Wei for commenting the research writing of Chap. 9.

The other one is the artificial intelligence method represented by artificial neural networks (ANN), support vector machines (SVM), and least square SVM (LSSVM). The existing studies demonstrate that the artificial intelligence method can capture the nonlinear characteristics of carbon price and obtain the good accuracy, while its performance is still limited to the stationarity of data.

In recent years, by decomposing a complex time series into a set of simple modes with simple structure, stationary fluctuation and strong regularity, multiscale ensemble prediction can significantly improve the accuracy for time series forecasting. Zhu (2012) combines empirical mode decomposition (EMD) with ANN to predict the carbon futures price in EUETS, and empirical results show that EMD plus ANN can outperform ANN remarkably. Zhang et al. (2015) develop a mixed model of ensemble EMD (EEMD), LSSVM, and GARCH to predict the WTI crude price in 2013, and discover that the mixed model can outmatch EEMD plus GARCH method, EEMD plus Particle Swarm Optimization (PSO)–LSSVM method, and PSO–LSSVM method. Yu et al. (2015) propose a novel decomposition-and-ensemble learning model integrating EEMD and extended extreme learning machine (EELM) for crude oil price forecasting, and empirical results demonstrate that the proposed model can statistically outperform the popular single model and similar decomposition–ensemble models. Therefore, multiscale ensemble prediction has become a new application prospect in the field of time series prediction and is expected to improve the prediction accuracy of carbon price.

Considerable amounts of achievements have been obtained by existing studies on carbon price prediction, which can provide a favorable reference for this research. However, two main drawbacks are found in existing studies: first, multiscale ensemble prediction is used to predict all the simple modes using the same model, rather than selecting an appropriate model for each mode according to its own data characteristics. Owing to the great differences among modes, the appropriate model is therefore required to be selected to predict each mode according to its own data characteristics (Zhang et al. 2008; Zhu et al. 2016). However, the existing methods are likely to limit the accuracy of carbon price prediction. Meanwhile, the typical EMD/EEMD is widely applied in multiscale ensemble prediction, which can result in the mode mixing (Hunag et al. 1999) and/or end effect (Xiong et al. 2014), affecting the decomposition quality of carbon price, and further influencing the accuracy of carbon price prediction. Second, although LSSVM is endowed with a favorable ability of nonlinear predictive modeling, this ability is limited by its kernel function type and model parameters (Rubio et al. 2011). Meanwhile, radical basis function (RBF) is predetermined by a majority of existing studies merely for the selection of model parameters excluding kernel function type (Zhu and Wei 2013; Chamkalani et al. 2014; Silva et al. 2015; Zhang et al. 2015). And few researches have been carried out to determine whether or not the selected kernel function is applicable to a specific problem, which can affect the accuracy of carbon price prediction.

This research aims to overcome the existing drawbacks of carbon price prediction, and develop a novel adaptive multiscale ensemble learning paradigm incorporating EEMD, PSO, and LSSVM with kernel function prototype to improve the accuracy of nonstationary and nonlinear carbon price time series forecasting. The innovation of this chapter mainly lies in two aspects: on one hand, it establishes a novel adaptive multiscale ensemble learning paradigm for carbon price forecasting. First, the extrema symmetry expansion EEMD is utilized to decompose the carbon price into simple modes, which can effectively restrain the mode mixing and end effects. Second, using the fine-to-coarse reconstruction algorithm, the high-frequency, low-frequency, and trend components are identified. Meanwhile, ARIMA is applicable to predicting the high-frequency components due to its strong ability of short-term memory. LSSVM, characterized by a favorable capture ability on nonlinear system, is therefore suitable for forecasting the low-frequency and trend components. At the same time, in order to take full use of the advantages of various kernel functions types and make up the drawbacks of single kernel function, a universal kernel function prototype is introduced, which can adaptively select the optimal kernel function type and model parameters according to the specific data using PSO. Finally, the prediction results of all the components acquired by different models are aggregated into the forecasting values of carbon price. On the other hand, compared with popular prediction methods, it is proved that the proposed model can effectively deal with the nonlinear and nonstationary carbon price. Besides, the proposed model can adaptively select the kernel function type and model parameters of LSSVM based on the data characteristics of each simple mode to build the optimal forecasting model, so as to improve the accuracy of carbon price prediction. Meanwhile, the proposed model can achieve the desired effects both from the perspectives of level and directional forecasting of carbon price.

9.2 Methodology

9.2.1 Kernel Function Prototype

Kernel function is a crucial element of LSSVM predictor. For a specific problem, the kernel function should capture the characteristics of this problem. However, it is hard to select an appropriate kernel function according to the prior characteristics of a specific problem. Researchers generally select the common kernel functions, such as the RBF kernel according to their experience, but the selected function possibly is not the optimal one for a specific problem. Therefore, how to select the optimal kernel function for a specific problem is required to be studied further.

The commonly used kernel functions include the linear kernel $K_{lin}(x_i, x_j) = (x_i, x_j)$, polynomial kernel $K_{poly}(x_i, x_j) = [(x_i, x_j) + t]^d$, RBF kernel $K_{rbf}(x_i, x_j) = \exp\left(-\frac{\|x_i - x_j\|^2}{2\sigma^2}\right)$, and Sigmoid kernel $K_{sig}(x_i, x_j) = \tan h[s \cdot (x_i, x_j) + h]$ (Smola 1998). According to the Mercer theory, a new kernel function can be obtained from combining various kernel functions linearly. Data contain the complete information of the specific problem. A kernel function, selected using the machine learning methods under the data-driven circumstance without any prior information, is proved to be the optimal one for a specific problem. Therefore, we introduce a new kernel function prototype into this chapter, which is defined as Eq. (9.1):

$$K(x,x') = \lambda_1 K_{sig}(x,x') + \lambda_2 K_{rbf}(x,x') + \lambda_3 K_{poly}(x,x') + \lambda_4 K_{lin}(x,x'). \qquad (9.1)$$

It can be found that kernel function prototype is a universal kernel function, which can not only generate the commonly used single kernel function, but also produce a new kernel function according to specific data. Thus, it is able to take full use of the advantages of different kernel functions to offset the drawbacks of single kernel function.

The model parameters to be determined for LSSVM with kernel function prototype can be divided into two categories: kernel function type parameter λ, and kernel function parameter u. Different combinations of λ and u are expected to create different kernel functions. When parameters λ and u are coded into the particles using the PSO algorithm, the optimal combination of λ and u for a specific problem can be adaptively selected. Afterward, the optimal kernel function for the problem is able to be adaptively determined after introducing the optimal combination of λ and u into the kernel function prototype. The obtained kernel function can be a single kernel function such as $\lambda_1 = 1$, $\lambda_i = 0$, $i = 2, 3, 4$, or a mixture kernel function such as $\lambda_i \neq 0$, $i = 1, 2, 3, 4$.

9.2.2 The Adaptive Parameter Selection for LSSVM with the PSO Algorithm

PSO is a swarm intelligent algorithm inspired by the foraging behavior of bird flocks. A potential solution of a real problem is called a particle in the PSO algorithm. First, the particles and their speeds are randomly initialized. The number of particles is called the population size, denoted as s, and the location of the i th particle in D–dimensional space is represented by $x_i = (x_{i1}, x_{i2}, \cdots, x_{iD})$, $i = 1, 2, \cdots, s$. The speeds of particles are indicated by $v_i = (v_{i1}, v_{i2}, \cdots, v_{iD})$, $i = 1, 2, \cdots, s$. D is the number of parameters to be optimized. The fitness value of each particle is calculated by the predefined fitness function. Afterward, according to the fitness value of each particle, the local optimum $p_{best} = (p_1, p_2, \cdots, p_D)$ and global optimum $G_{best} = (G_1, G_2, \cdots, G_D)$ of each particle are updated. Finally, the local and global optimums of each particle are tracked dynamically by formulas Eqs. (9.2) and (9.3), so as to update its speed and location:

$$v_{ij}(t+1) = w(t) \cdot v_i(t) + c_1 \cdot r_1 \cdot [P_j(t) - x_{ij}(t)] + c_2 \cdot r_2 \cdot [G_j(t) - x_{ij}(t)], \qquad (9.2)$$

$$x_{ij}(t+1) = x_{ij}(t) + v_{ij}(t+1), \qquad (9.3)$$

where $j = 1, 2, \cdots, D$. t is the current iteration number. r_1, r_2 are the random numbers distributed uniformly in interval of $(0,1)$, respectively. c_1, c_2 are the acceleration factors. The inertia weight $w(t)$ is calculated using formula Eq. (9.4):

$$w(t) = w_{max} - \frac{w_{max} - w_{min}}{t_{max}} \times t, \tag{9.4}$$

where w_{max} and w_{min} are, respectively, the initial and final inertial weights.

In the updating process, the speed of each particle is limited into a preset interval of $[-V_{max}, V_{max}]$. If $v_{ij}(t+1) > v_{max}$, then $v_{ij}(t+1) = v_{max}$. If $v_{ij}(t+1) < -v_{max}$, then $v_{ij}(t+1) = -v_{max}$. The termination condition for the iteration of the PSO algorithm is to reach the preset maximum iteration number.

As a class of time series forecasting problem in essence, carbon price prediction is based on the phase space reconstruction (PSR). The quality of PSR can directly affect the establishment and prediction of the follow-up models. PSR is involved in the selections of embedding dimension (m) and delay (τ). Too small m cannot display the fine structure of chaotic system for carbon price, while too large m will complicate the computation and therefore can cause noise. Likewise, if τ is too small, the adjacent delay coordinate elements differ slightly in the phase space and therefore can lead to information redundancy, whereas with an overlarge τ, the adjacent delay coordinate elements are not associated, which can result into information lose and thus can fold signal trajectories. Therefore, m and τ can directly influence the accuracy of carbon price forecasting.

The carbon price system is time varying: as the newly input and output data are obtained, the system state also changes constantly. Therefore, in order to have the model accurately reflect the state of current system, new data are required by the model, while the old data that have little correlation with current state may be ignored or its contribution is reduced. Meanwhile, based on the principle that the closer the data to those to be predicted, the larger the influence is, and vice versa, it can be concluded that the data close to the prediction have a much more influence on the prediction results. Therefore, a modeling data interval that can slide with time is required to be established, namely, the sliding time window tw. The newly obtained information of time series is dynamically introduced into the model to capture the time structure of the carbon price data.

Therefore, the model parameters to be adaptively selected by the PSO algorithm include kernel function type parameter λ, kernel function parameter μ, PSR m and τ, sliding time window tw, and penalty factor γ. These parameters are coded into the particles through the PSO algorithm to conduct adaptively the optimization selection, so as to simultaneously determine the optimal parameter combination for a LSSVM predictor according to the data characteristics. The adaptive model selection algorithm, called PSO–LSSVM, is proposed in this chapter, as illustrated in Fig. 9.1.

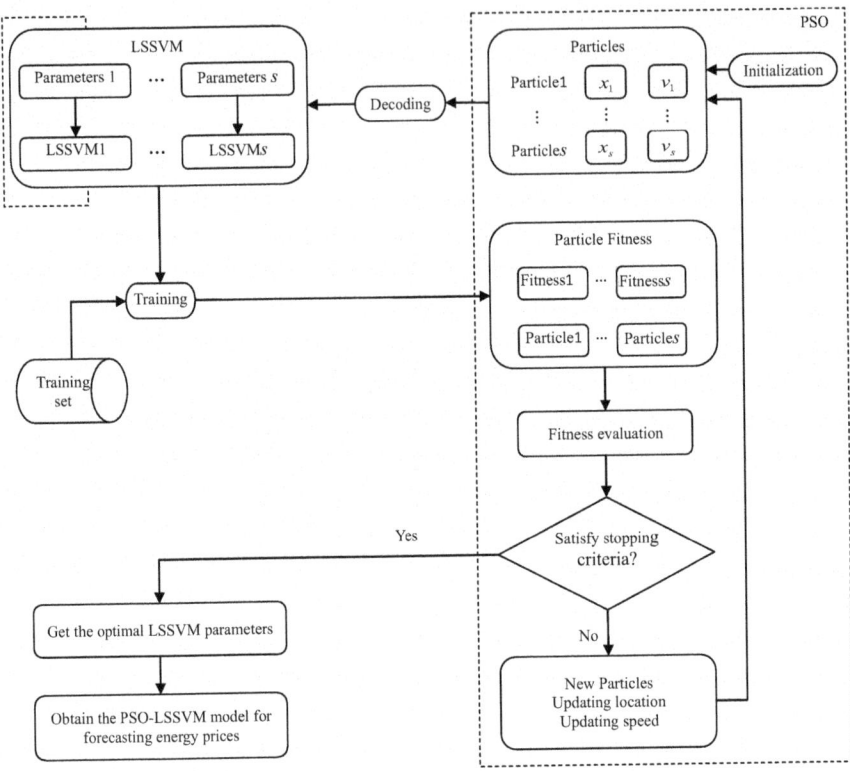

Fig. 9.1 The process of the adaptive PSO–LSSVM model

Step 1: *Coding.* The integer coding is used for m, τ, d and tw, and the real coding is used for others. All the model parameters are coded into a particle, $x_i = (x_{i1}, x_{i2}, \ldots, x_{i13})$.

Step 2: *Initializing.* The particles and their speeds are generated randomly, and the maximum speed, iteration number, and range of inertia weight are set here.

Step 3: *Defining the fitness function of each particle.* The minimization of the root mean square error (RMSE) of the training set is defined as the objective function:

$$\min RMSE(\lambda, u, m, \tau, tw, \gamma) = \sqrt{\frac{1}{n}\sum_{i=1}^{n}\left[y_i - \varphi(x_i, \lambda, u, m, \tau, tw, \gamma)\right]^2},$$

where $\varphi(\cdot)$ is a nonlinear mapping function determined by the LSSVM predictor with given parameters and training samples. x_i is the samples of training set. y_i is the target value corresponding to the training sample x_i, and n is the size of training set. In this chapter, the fitness function of each particle is defined by the RMSE as Eq. (9.5):

$$F_{\text{fitness}} = RMSE(\lambda, u, m, \tau, tw, \gamma). \tag{9.5}$$

Step 4: *Calculating the fitness value of each particle.* The fitness value of each particle, $F(P_{\text{ipresent}})$, is calculated using the fitness function. The local optimum of each particle P_{ibest} is set as the current location, and afterward, the particle with the optimal fitness value is selected as the initial global optimum G_{best}.

Step 5: *Updating the speed, location, and fitness value of each particle.* The updates are performed by formulas (1), (2), (3), and (4).

Step 6: Comparing $F(P_{\text{ipresent}})$ with $F(P_{\text{ibest}})$, if $F(P_{\text{ipresent}}) < F(P_{\text{ibest}})$, then $P_{\text{ibest}} = P_{\text{ipresent}}$.

Step 7: Comparing the updated $F(P_{\text{ipresent}})$ and $F(G_{\text{best}})$, if $F(P_{\text{ipresent}}) < F(G_{\text{best}})$, then $G_{\text{best}} = G_{\text{ipresent}}$.

Step 8: Judging whether or not the termination condition is satisfied. If it is, stop the search process, and output the optimized LSSVM parameters. At this point, the best LSSVM predictor is obtained. Otherwise, let $t = t + 1$, and return to step 4.

9.2.3   The Proposed Adaptive Multiscale Ensemble Model for Carbon Price Forecasting

The proposed adaptive multiscale ensemble learning paradigm for nonstationary and nonlinear carbon price forecasting, namely, EEMD-HLT-Σ, which integrated EEMD, kernel function prototype, LSSVM and ARIMA organically, can be divided into four stages, as shown in Fig. 9.2.

Stage 1: Decomposition of EEMD

The carbon price is decomposed into m independent IMFs with different amplitudes and frequencies and one residue from high frequency to low frequency using the extrema symmetry expansion EEMD. These independent IMFs are characterized by simple structure, stable fluctuation, and strong regularity as formula Eq. (9.6), which can be predicted readily:

$$X(t) = \sum_{j=1}^{m} c_j(t) + r(t) = HFs + LFs + T. \tag{9.6}$$

Based on the frequencies of IMFs, these IMFs can further be divided into three components: (i) High-frequency components (HFs), characterized by high frequency and low amplitude, can reflect the characteristics of random fluctuations of carbon price induced by the supply and demand imbalance of short-term market in normal operation. Although HFs fluctuate frequently in short-term, no long-term effect is generated by themselves. Meanwhile, the fast-changing price also

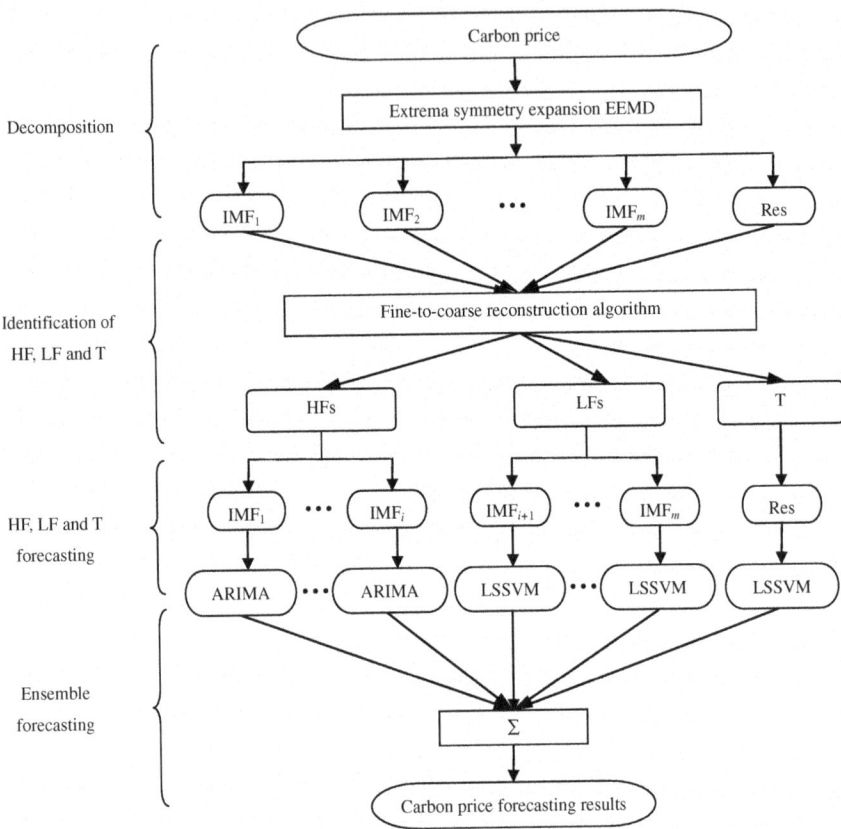

Fig. 9.2 The process of EEMD-HLT-Σ model for forecasting carbon price

accompanies with fast recovery speed. The rise (decline) of price always follows closely the decline (rise) of price, that is, the gathering effects of fluctuations. (ii) Low-frequency components (LFs) are characterized by low frequency and high amplitude, indicating the periodic fluctuations of carbon price influenced by outside heterogeneous environments. The rapid rise or fall of LFs is a demonstration of the influences of major unexpected events. Though the fluctuation frequencies of LFs are low, they may affect the carbon price greatly and even change the pricing mechanism of energy market. (iii) Trend component (T) is represented by the slow change near the long-term mean value, which can describe the stable change of carbon price in long term. Hence, each component is endowed with unique characteristics. Therefore, the accuracy of carbon price prediction is bound to be improved once the appropriate forecasting model for each component is selected according to its data characteristics.

Stage 2: Identification of HFs, LFs, and T

The different data characteristics exhibited by HFs, LFs, and T, respectively, show the different intrinsic properties of initial carbon price. Once the decomposition of EEMD is finished, the fine-to-coarse reconstruction algorithm (Zhang et al. 2008) is employed to identify the HFs, LFs, and T, so as to select the optimal models for forecasting each component. The identification process is given as follows:

Step 1: Calculating the mean value $\bar{s}_i$ of the superposed sum series $s_i = \sum_{k=1}^{i} IMF_k (i = 1, 2, \ldots, m)$ from IMF_1 to IMF_i.

Step 2: Selecting the significance level α, usually $\alpha = 0.05$. The t test is conducted to judge whether or not the mean value $\bar{s}_i$ deviates from 0 significantly.

Step 3: If $\bar{s}_i$ starts to deviate from 0 significantly at i, the IMFs from IMF_i to IMF_m are identified as LFs, and the rest IMFs are HFs. Meanwhile, the residue is identified as T.

Stage 3: Prediction of HFs, LFs, and T

The following strategy is adopted to predict carbon price: first, according to the data characteristics of each component, the appropriate model is selected to predict each component alone, and then the prediction results of all the components are aggregated into the forecasting value of the original carbon price.

The forecasting model for time series $X_t (t = 1, 2, \ldots, n)$ can be generally described as Eq. (9.7):

$$\hat{X}_{t+h} = f(X_t, X_{t-\tau}, X_{t-2\tau}, \ldots, X_{t-(m-1)\tau}) + \varepsilon_t, \tag{9.7}$$

where $\hat{X}_t$ is the predicted value of initial time series, h is the prediction horizon, m, τ are the embedding dimension and delay of PSR, respectively, and ε_t is the perdition errors. When $h = 1$, it is one-step-ahead forecasting. When $h \geq 2$, it is h-step-ahead forecasting.

1. *Prediction of HFs.* HFs demonstrate the short-term fluctuations of carbon price, and is endowed with high randomness. The econometric model with a strong capacity of short-term memory, and a better modeling ability on a random process, such as ARIMA, is applicable to forecasting HFs. As to the specific modeling processes, the method provided by Box and Jenkins in (1976) can help identify the model order and establish ARIMA model. ARIMA model is the combination of AR model and MA model, characterized by high prediction efficiency and high prediction accuracy.

2. *Forecasting of LFs.* Because of the strong periodic fluctuations of LFs, LSSVM, which has a favorable modeling performance on nonlinear time series, is employed to predict the LFs. The kernel function prototype is selected as the kernel function of LSSVM model. According to the data characteristics of each

LF, the proposed method can adaptively select the optimal kernel function type and model parameters mentioned in Sect. 9.2.2.

3. *Forecasting of T.* The changing direction of trend component is clear and stable, and therefore, it can be predicted by LSSVM. The model selection of LSSVM for Ts is same as that of the prediction of LFs.

Stage 4: Integration of the prediction results of HFs, LFs, and T

The prediction results of HFs, LFs, and T are superposed to obtain the predicted values of the initial carbon price as formula Eq. (9.8):

$$\hat{X} = \hat{H}Fs + \hat{L}Fs + \hat{T},   \tag{9.8}$$

where $\hat{X}$ is the predicted value of initial carbon price, and $\hat{H}Fs$, $\hat{L}Fs$, and $\hat{T}$ are the predicted values of HFs, LFs, and T, respectively.

9.3 Empirical Analysis

9.3.1 Data

In order to verify the validity of the proposed model, this chapter selects the daily European Union Allowance futures price (DEC 15) matured in December 2015 from the intercontinental exchange (ICE) which has the maximum trading volume under the EU ETS, as an investigation sample. For the convenience of predictive modeling, the sample is divided into two subsets: the training set and the testing set. The training set is used to establish prediction models, and the testing set is employed to verify the validity of the established models. The divided samples of carbon price are reported in Table 9.1.

9.3.2 Evaluation Criteria

The RMSE and directional prediction statistics (D_{stat}) are applied as the evaluation criteria to the prediction performance of the established models. For comparing that whether or not the prediction accuracy of model A is obviously better than that of

Table 9.1 Samples of carbon price

Carbon price		Size	Date
DEC 15	Sample set	861	November 29, 2011–April 14, 2015
	Training set	701	November 29, 2011–August 26, 2014
	Testing set	160	August 27, 2014–April 14, 2015

model B, we introduce the Diebold–Mariano (DM) test, Pesaran–Timmerman (PT) test, and rate test (RT) into study.

The statistics of PT test is defined as $z_{PT} = \dfrac{\hat{p}-p^*}{\sqrt{p^*(1-p^*)/n}} \sim N(0,1), n \to \infty$, where $\hat{p} = \frac{1}{n}\sum_{t=1}^{n} H_t[(x_{t+1}-x_t)(\hat{x}_{t+1}-x_t)]$, namely, the accuracy of directional prediction of the model, measured by D_{stat}. Meanwhile, $p^* = p_1\hat{p}_1 + (1-p_1)(1-\hat{p}_1)$, $p_1 = \frac{1}{n}\sum_{t=1}^{n} H_t(x_{t+1}-x_t)$, $\hat{p}_1 = \frac{1}{n}\sum_{t=1}^{n} H_t(\hat{x}_{t+1}-x_t)$, and $H(x) = \begin{cases} 1, & x \geq 0 \\ 0, & x < 0 \end{cases}$.
The null assumption of PT test is that the predicted direction and the actual direction of the model are independent with each other. When the absolute value of z_{PT} is greater than 1.96 in the two-sided test, the null hypothesis is rejected at the significance level of 5%.

9.3.3 Nonstationary and Nonlinear Tests of Carbon Price

Augmented Dicky–Fuller (ADF) test is an effective method to check the stationarity of a time series, while Brock–Dechert–Scheinkman (BDS) test can effectively find out the nonlinear characteristics of a time series, where the embedding dimension is set to 2–5, and the dimensional distance is set to 0.7 times of the variance of data. The EViews 6.0 software package developed by Quantitative Micro Software Corporation is employed to test the nonstationarity and nonlinearity of carbon prices, and the test results are shown in Tables 9.2 and 9.3. As observed in the two tables, ADF test demonstrated that the p value of carbon prices are no less than 10%, indicating that carbon prices at the significance level of 10% are nonstationary. BDS test demonstrated that all the p values of carbon prices are 0, which shows that carbon prices are nonlinear at the significance level of 1%.

Table 9.2 ADF test results

Carbon price	t-Stat	Prob.	Stationary
DEC 15	−2.5045	0.1147	×

Note × shows that carbon price is nonstationary at the significance level of 10%

Table 9.3 BDS test results

Carbon price	Embedding dimension								
	2		3		4		5		Linearity
	Stat.	Prob.	Stat.	Prob.	Stat.	Prob.	Stat.	Prob.	
DEC 15	0.018	0.000	0.043	0.000	0.069	0.000	0.086	0.000	×

Note × shows that carbon price is nonlinear at the significance level of 1%

9.3.4 Decomposition of EEMD

EEMD is applied to decompose the carbon prices, in which N is set as 100, and σ is 0.2 times standard deviation of each carbon price. The termination condition is to reach the maximum shifting times of 10. Because end effect is likely to be generated by EEMD, the extrema symmetry expansion EEMD and envelope extrema expansion EEMD are utilized to restrain the happening of end effect and model mixing.

The decomposition results of EEMD are orthogonal, thus the sum of energies of all components is equal to that of the initial time series. However, because of end effect, some illusive components may be generated by EEMD, resulting in the increased sum of energies of all components (Wu and Huang 2009; Ren et al. 2012). Therefore, the degree of energies changes before and after the time series decomposed by EEMD can reflect the influence of end effect. The energy of a time series is defined as Eq. (9.9):

$$E = \sqrt{\frac{1}{n} \sum_{i=1}^{n} s^2(i)}, \tag{9.9}$$

where E is the energy of a time series, $s(i)$ is the real value of time series, and n is the size of time series.

The evaluation index θ is defined as Eq. (9.10):

$$\theta = \frac{\left| \sqrt{\sum_{i=1}^{m+1} E_i^2} - E_{X(t)} \right|}{E_{X(t)}}, \tag{9.10}$$

where $E_{X(t)}$ is the energy value of the initial time series, E_i is the energy value of the ith IMF, and m is the number of IMFs.

It can be found that $\theta \geq 0$. If $\theta = 0$, it indicates that no end effect is generated by EEMD. The greater the θ, the greater the influence of end effect. Table 9.4 illustrates the values of θ obtained from the extrema symmetry expansion EEMD and envelope extrema expansion EEMD. It is discovered that the value of θ for the extrema symmetry expansion EEMD is turned out to be the minimum in all the time series. Therefore, the extrema symmetry expansion EEMD is employed to decompose the carbon prices in this chapter, and the numbers of obtained IMFs are,

Table 9.4 Comparisons of various expansion EEMD algorithms

Carbon price	θ		
	EEMD	envelope extrema expansion EEMD	extrema symmetry expansion EEMD
DEC 15	0.0129	0.0108	0.0084

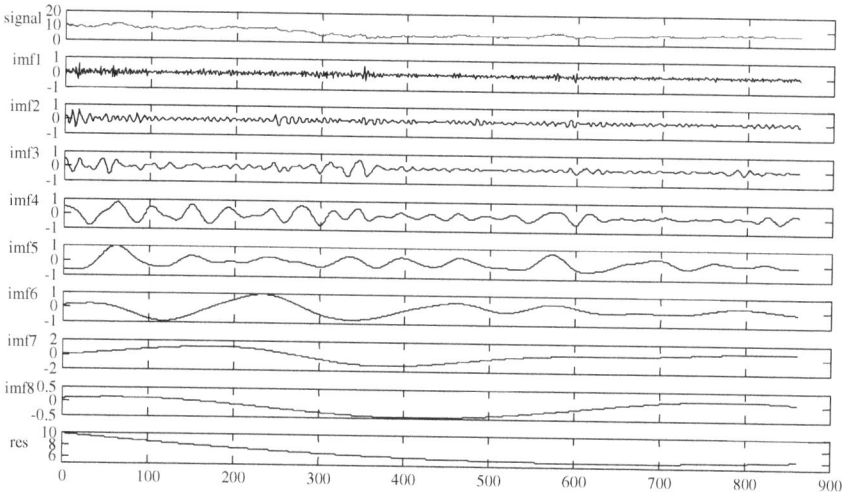

Fig. 9.3 Decomposition results of DEC 15

respectively, 8, along with one residue, as shown in Fig. 9.3. All the IMFs in Fig. 9.3 are arranged from high frequency to low frequency, which exhibit different frequencies and amplitudes, and the last one is the residue. It can also be discovered that HFs are characterized by randomness, while the LFs are with strong periodic fluctuations, and the residue shows its tendency. Compared with each initial carbon prices, these components are obviously endowed with characteristics of simpler structure, more stationary fluctuation, and greater regularity, and are more easily to be predicted.

9.3.5 Identification of HFs, LFs, and T

The fine-to-coarse reconstruction method is applied on IMF_1–IMF_m to calculate the test values of t and p values of $\bar{s}_i$ from fine scale to coarse scale, and the results are listed in Table 9.5. It is discovered that at the significance level of 5%, $\bar{s}_i$ of DEC 15 starts to significantly deviate from 0 at the point $i = 7$, and therefore, IMF_1–IMF_6 belong to HFs, while IMF_7–IMF_8 are the LFs. The residue is identified as the T.

Table 9.5 Identification results of HFs and LFs

Carbon price	Parameters	S_1	S_2	S_3	S_4	S_5	S_6	S_7	S_8
DEC 15	$\bar{s}_i$	−0.0011	−0.0013	0.0005	0.0043	−0.012	−0.0313	0.1291	0.0108
	t	−0.3367	−0.2217	0.0561	0.3001	−0.6090	−1.2256	3.4464	0.2555
	p value	0.7365	0.8246	0.9553	0.7642	0.5427	0.2207	0.0006	0.7984

9.3.6   Forecasting Results and Analysis

The HFs, LFs, and T are identified first in the process of one-step-ahead forecasting carbon prices using the proposed EEMD-HLT-Σ model. Afterward, the HFs are predicted through the ARIMA modeling, and the LFs and T are forecasted utilizing the PSO–LSSVM model. Finally, the prediction results of all components are aggregated into prediction values of the initial carbon prices. For comparisons, the single ARIMA and LSSVM models and multiscale ensemble prediction models EEMD-ARIMA-Σ (summarized the values of all IMFs and residue predicted by ARIMA) and EEMD-LSSVM-Σ (summarized the values of all IMFs and residue predicted by LSSVM) are applied for carbon prices forecasting. Meanwhile, all ARIMA models are established using the EViews 6.0 software package. In addition, the kernel function prototype is selected for all LSSVM models, and these models are realized through the programming of Matlab R2012b platform developed by the MathWorks, Inc. The numebers of particles and maximum iteration are, respectively, as 100 and 50. Meanwhile, $m \in [1, 10], \tau \in [1, 4], tw \in [50, 400],$ $d \in [0, 10], t \in [0, 20], s \in [0.01, 50], h \in [-100, 100], \lambda_i \in [-50, 50], i = 1, 2, 3, 4,$ $\gamma \in [0.01, 100000], \sigma^2 \in [0.01, 100000], c_1 = c_2 = 2, w \in [0.1, 0.9],$ and $v_{\max} = 50.$ The fitness function is defined as $F_{\text{fitness}} = RMSE(\lambda, u, m, \tau, w, \gamma).$ Table 9.6 illustrates the optimal parameters of proposed EEMD-HLT-Σ model for carbon prices forecasting. The comparisons between the RMSE and D_{stat} of each prediction model are revealed in Table 9.7. The results of DM test, PT test, and RT test are, respectively, shown in Tables 9.8 and 9.9. Figure 9.4 demonstrates the out-of-sample prediction results of each carbon price using EEMD-HLT-Σ model.

In terms of level forecasting as demonstrated by RMSE, we can find out that, first, all the multiscale ensemble prediction models including EEMD-ARIMA-Σ, EEMD-LSSVM-Σ, and EEMD-HLT-Σ obviously outperform each single prediction model such as ARIMA and LSSVM. This is mainly attributed to the fact that the decomposition of EEMD can significantly improve the prediction ability of the model. Second, as for the single prediction model, the prediction accuracy of LSSVM model is superior to that of ARIMA model because the former is nonlinear while the former is linear. Due to the great fluctuations, nonstationarity, and nonlinearity of carbon prices, the linear model is therefore not suitable for forecasting carbon prices. Meanwhile, based on kernel function prototype, driven by the data characteristics and the adaptive selections on kernel function type and model parameters by the global optimization of PSO, the learning and prediction abilities of LSSVM are greatly improved. Among the multiscale ensemble prediction models, the EEMD-LSSVM-Σ model displays higher prediction accuracies on the prices of crude oil than that of EEMD-ARIMA-Σ model. However, with respect to the prediction accuracy on the prices of carbon, electricity, natural gas, and gasoline, the RMSE of EEMD-ARIMA-Σ model is basically the same with that of EEMD-LSSVM-Σ model. The main reason may lie in the fact that carbon prices decomposed by EEMD are endowed with simpler structure and more stationary fluctuations, and are more likely to be forecasted. As a result, both the ARIMA and

Table 9.6 The optimal parameters for carbon price prediction using EEMD-HLT-Σ model

Carbon price	Parameters	Original series	IMF1	IMF2	IMF3	IMF4	IMF5	IMF6	IMF7	IMF8	Residue
DEC 15	d	1	ARIMA (4, 0, 2)	ARIMA (4, 0, 5)	ARIMA (2,0, 0)	ARIMA (4, 0, 0)	ARIMA (4, 0, 0)	ARIMA (4, 0, 0)	7	10	0
	t	3.35							20	4.21	1.56
	σ^2	6658.94							10000	0.01	1056.3
	s	41.99							0.01	10.26	12.15
	h	42.31							−4	−100	−78.06
	λ_1	4.05							−20	2.44	−2.45
	λ_2	17.95							20	−10	1.49
	λ_3	10.07							−20	−7.92	−1.66
	λ_4	-8.47							−20	−10	6.36
	γ	4933.33							0.01	5172.55	3380.98
	m	4							10	10	7
	τ	1							4	4	1
	w	384							400	400	236

te ARIMA(p, d, q) model. p, q are, respectively, the orders of autoregressive and moving average, and d is the degree of ordinary differencing

Models	RMSE	D_{stat}
	DEC 15	
ARIMA	0.111	0.65
LSSVM	0.105	0.669
EEMD-ARIMA-Σ	0.052	0.894
EEMD-LSSVM-Σ	0.052	0.906
EEMD-HLT-Σ	0.047	0.919

Table 9.7 Comparison of RMSE and D_{stat} of each prediction model

LSSVM models can achieve satisfying prediction results. Compared with other models including single model and multiscale ensemble models, the proposed EEMD-HLT-Σ model remains the maximum accuracy of level forecasting on all carbon prices. This is possibly because through EEMD, the proposed model first identifies the HFs, LFs, and T according to their own different data characteristics. Then, the more appropriate ARIMA and LSSVMM models are used respectively to forecast the HFs, LFs, and T, and thus the prediction ability of the proposed model is enhanced.

According to the DM test results, the following conclusions can be obtained: first, all the multiscale ensemble models precede the single prediction model remarkably at the significance level of 1%, which indicates that the principle of decomposition–ensemble can significantly improve the prediction ability of model. Second, at the significance level of 5%, the EEMD-LSSVM-Σ model is superior to EEMD-ARIMA-Σ model on crude oil price; however, there exists no significant difference between the two models when it comes to the other carbon prices. This is mainly because the decomposition of EEMD makes the data more stationary, regular and likely to be predicted. Third, at the significance level of 10%, the EEMD-HLT-Σ model notably outperforms other prediction models on the prices of carbon, gasoline, and crude oil, while there is no significant difference with the EEMD-LSSVM-Σ model on electricity price, as well as with EEMD-LSSVM-Σ and EEMD-ARIMA-Σ models on the gas price.

From the perspective of directional prediction, the conclusions similar to those of RMSE can be obtained according to the results of D_{stat}: first, the multiscale ensemble prediction models including EEMD-ARIMA-Σ, EEMD-LSSVM-Σ, and EEMD-HLT-Σ significantly excel all the single prediction model such as ARIMA and LSSVM. Second, among single prediction model, LSSVM model has an obviously superior to ARIMA model; while for multiscale ensemble prediction models, the prediction accuracy of EEMD-LSSVM-Σ model is superior to that of EEMD-ARIMA-Σ model on the carbon prices except the gas price, where the former is slightly inferior to the latter. Third, the prediction accuracy of the proposed EEMD-HLT-Σ model is slightly inferior to that of the EEMD-LSSVM-Σ on the electricity and gasoline prices, but the former is always superior to other models of the two prices. Meanwhile, the prediction accuracy of EEMD-HLT-Σ model outperforms all the models for predicting carbon prices except the electricity and gasoline prices.

Table 9.8 Comparison of DM and *RT* of each prediction model

DEC 15	DM benchmark				RT benchmark			
	ARIMA	LSSVM	EEMD-ARIMA-Σ	EEMD-LSSVM-Σ	ARIMA	LSSVM	EEMD-ARIMA-Σ	EEMD-LSSVM-Σ
LSSVM	2.315 (0.011)				0.196 (0.422)			
EEMD-ARIMA-Σ	6.302 (0.000)	5.671 (0.000)			2.875 (0.002)	2.694 (0.004)		
EEMD-LSSVM-Σ	6.465 (0.000)	5.827 (0.000)	0.065 (0.474)		3.053 (0.001)	2.874 (0.002)	0.206 (0.418)	
EEMD-HLT-Σ	6.693 (0.000)	6.081 (0.000)	3.157 (0.001)	2.276 (0.012)	3.235 (0.001)	3.058 (0.001)	0.425 (0.672)	0.219 (0.827)

Note $^{*}z_{DM}$(p value) and $^{**}z_{RT}$(p value)

Table 9.9 Comparisons of PT of each prediction model

Carbon price	Parameters	ARIMA	LSSVM	EEMD-ARIMA-Σ	EEMD-LSSVM-Σ	EEMD-HLT-Σ
DEC 15	t value	3.85	4.364	10.048	10.333	10.633
	p value	0.000	0.000	0.000	0.000	0.000

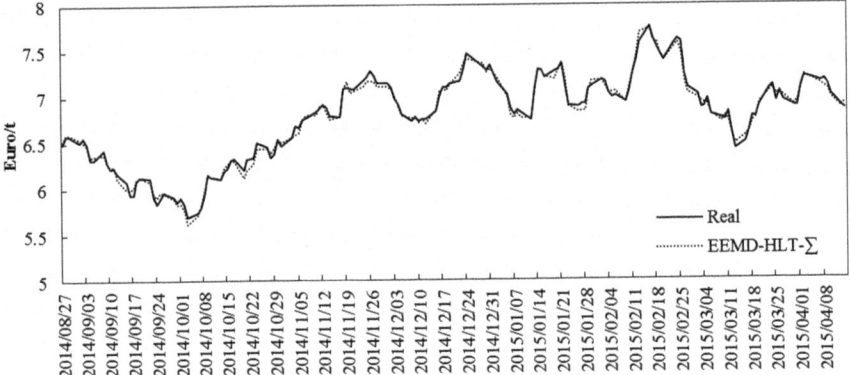

Fig. 9.4 Out-of-sample prediction results of DEC 15 using EEMD-HLT-Σ model

PT test demonstrates that the multiscale ensemble models can obtain high accuracy of directional prediction at the confidence level of 99% for all carbon prices. The RT test further reveals the results of PT test that multiscale ensemble prediction models can notably precede single prediction model at confidence level of 99%. Meanwhile, the LSSVM model is superior to ARIMA model, but this advantage is only significant for predicting the prices of gasoline and crude oil at the significance level of 10%. Compared with all the single prediction model, EEMD-HLT-Σ is endowed with obvious superiority at the confidence level of 99%, but its superiority is not always significant when it is compared with other multiscale ensemble prediction models.

The out-of-sample prediction results on the testing set of carbon prices revealed that the proposed EEMD-HLT-Σ model can acquire a favorable prediction effect both in the level and directional predictions. Therefore, the out-of-sample predicted values are very close to their own actual values.

It is worth noting that the RMSE of EEMD-HLT-Σ model is superior to other models, but the values of D_{stat} of electricity and gasoline prices for EEMD-HLT-Σ model are slightly lower than those of EEMD-LSSVM-Σ model. Similarly, the values of D_{stat} of electricity, natural gas, and gasoline prices for EEMD-LSSVM-Σ model are slightly lower than those of EEMD-ARIMA-Σ model. This indicates that the model with high accuracy of level forecasting is not necessarily endowed with high accuracy of directional prediction.

Above all, according to the empirical analysis on the prices of five kinds of energies, the following conclusions can be obtained: first, the prediction accuracy of the proposed EEMD-HLT-Σ model is generally higher than those of other models both in the level and directional predictions. Its accuracy of level forecasting is higher than those of other models particularly. Second, multiscale ensemble prediction models can outperform single prediction model, which suggests that the strategy of decomposition–ensemble can significantly improve the prediction ability of the model on carbon prices. Third, due to the high nonstationarity and nonlinearity of carbon prices fluctuations, nonlinear model is more applicable to forecast carbon prices than linear model. Finally, the proposed EEMD-HLT-Σ model can obviously improve the prediction accuracy on carbon prices, and therefore can be considerable as a competitive method for carbon price prediction.

9.4 Conclusion

As for the high complex, nonstationary, and nonlinear carbon price, a new adaptive multiscale ensemble learning paradigm incorporating EEMD, PSO, and LSSVM with kernel function prototype is put forward for carbon price prediction in this chapter. First, the extrema symmetry expansion EEMD algorithm is employed to decompose the carbon prices into a set of IMFs and one residue, which are with strong regularity, simple structure, and smooth fluctuations, and are likely to be predicted. Afterward, HFs, LFs, and T are identified using the fine-to-coarse reconstruction algorithm. Later, the ARIMA model is utilized to predict HFs, while the adaptive PSO–LSSVM models with kernel function prototype are used for forecasting LFs and T according to their own data characteristics. Finally, the prediction results of HFs, LFs, and T obtained by different models are aggregated into the prediction values of the initial carbon prices time series. As demonstrated by the empirical results, compared with the common popular prediction methods, the proposed adaptive multiscale ensemble learning paradigm can significantly improve the prediction accuracy of carbon prices, with high accuracies both in the level and directional predictions. This suggests that the proposed model is a promising and competitive approach for predicting the high nonstationary, nonlinear, and irregular carbon prices time series. In the future, one of our work is to investigate that how to establish the optimal prediction model for each component based on data characteristics of each component along with the addition of more influential factors in the modeling process. By doing so, it is expected to further improve the prediction accuracy of the high nonstationary, nonlinear, and irregular carbon prices time series.

References

Anbazhagan S, Kumarappan N (2014) Day-ahead deregulated electricity market price forecasting using neural network input featured by DCT. Energy Convers Manage 78(2):711–719

Bastianin A, Galeotti M, Manera M (2014) Forecasting the oil–gasoline price relationship: do asymmetries help? Energy Econ 46:S44–S56

Box GEP, Jenkins GM (1976) Time series analysis: forecasting and control. Holden Day, San Francisco

Byun SJ, Cho H (2013) Forecasting carbon futures volatility using GARCH models with energy volatilities. Energy Econ 40:207–221

Chamkalani A, Zendehboudi S, Bahadori A et al (2014) Integration of LSSVM technique with PSO to determine asphaltene deposition. J Petrol Sci Eng 124:243–253

Chiroma H, Abdul-Kareem S, Abubakar A, Zeki AM, Usman MJ (2014) Orthogonal wavelet support vector machine for predicting crude oil prices. Lect Notes in Electr Eng 285:193–201

Diebold FX, Mariano RS (1995) Comparing predictive accuracy. J Bus Econ Stat 13(3):253–263

García-Martos C, Rodríguez J, Sánchez MJ (2013) Modelling and forecasting fossil fuels, CO_2 and electricity prices and their volatilities. Appl Energy 101:363–375

Hou A, Suardi S (2012) A nonparametric GARCH model of crude oil price return volatility. Energy Econ 34:618–626

Huang NE, Shen Z, Long SR (1998) The empirical mode decomposition and the Hilbert spectrum for non-linear and non-stationary time series analysis. Proc Roy Soc London 454:903–995

Hunag NE, Shen Z, Long SR (1999) A new view of nonlinear water waves: the Hilbert spectrum. Annu Rev Fluid Mech 31:417–457

Kennedy J, Eberhart RC (1995) Particle swarm optimization. Proceedings IEEE Conference on Neural Networks. Perth: Piscataway 4: 1942–1948

Pesaran MH, Timmermann P (1992) A simple nonparametric test of predictive performance. J Bus Econ Stat 10:461–465

Ren DQ, Yang SX, Wu ZT, Yang B (2012) Research on end effect of LMD based time-frequency analysis in rotating machinery fault diagnosis. China Mech Eng 23(8):951–956

Rubio G, Pomares H, Rojas I, Herrera LJ (2011) A heuristic method for parameter selection in LS-SVM: application to time series prediction. Int J Forecast 27(3):725–739

Salehnia N, Falahi MA, Seifi A, Adeli MHM (2013) Forecasting natural gas spot prices with nonlinear modeling using Gamma test analysis. J Nat Gas Sci Eng 14:238–249

Shu ZP, Yang ZC (2006) A better method for effectively suppressing end effect of empirical mode decomposition. J North Western Poly Tech Univ 24(5):639–643

Silva DA, Silva JP, Neto ARR (2015) Novel approaches using evolutionary computation for sparse least square support vector machines. Neurocomputing 168:908–916

Suykenns JAK, Vandewalle J (1999) Least squares support vector machine. Neural Process Lett 9(3):293–300

Smola AJ (1998) Learing with kernels. Ph.D. thesis, Department of Computer Science. Technical University Berlin, Germany

Wu Z, Huang NE (2009) Ensemble empirical mode decomposition: a noise-assisted data analysis method. Adv Adapt Data Anal 1(1):1–41

Xiong T, Bao YK, Hu ZY (2014) Does restraining end effect matter in EMD-based modeling frame work for time series prediction? Some experimental evidences. Neurocomputing 123:174–184

Yu L, Dai W, Tang L (2015) A novel decomposition ensemble model with extended extreme learning machine for crude oil price forecasting. Eng Appl Artif Intell 47:110–121

Zhang X, Lai KK, Wang SY (2008) A new approach for crude oil price analysis based on empirical mode decomposition. Energy Econ 30:905–918

Zhang W, Niu P, Li G, Li P (2013) Forecasting of turbine heat rate with online least squares support vector machine based on gravitational search algorithm. Knowl-Based Syst 39:34–44

Zhang JL, Zhang YJ, Zhang L (2015) A novel hybrid method for crude oil price forecasting. Energy Econ 49:649–659

Zhou LG, Lai KK, Yu L (2009) Credit scoring using support vector machines with direct search for parameters selection. Soft Comput 13:149–155

Zhu BZ (2012) A novel multiscale ensemble carbon price prediction model integrating empirical mode decomposition, genetic algorithm and artificial neural network. Energies 5(12):355–370

Zhu BZ, Wei YM (2013) Carbon price prediction with a hybrid ARIMA and least squares support vector machines methodology. Omega 41:517–524

Zhu BZ, Ma SJ, Chevallier J, Wei YM (2014) Examining the structural changes of European carbon futures price 2005–2012. Appl Econ Lett 21:1381–1388

Zhu B, Shi X, Chevallier J et al (2016) An adaptive multiscale ensemble learning paradigm for nonstationary and nonlinear energy price time series forecasting. J Forecast 35(7):633–651

Ziel F, Steinert R, Husmann S (2015) Efficient modeling and forecasting of electricity spot prices. Energy Econ 47:98–111

Index

© Springer International Publishing AG 2017
B. Zhu and J. Chevallier, *Pricing and Forecasting Carbon Markets,*
DOI 10.1007/978-3-319-57618-3

The manufacturer's authorised representative in the EU is Springer
Nature Customer Service Centre GmbH, Europaplatz 3, 69115 Heidelberg,
Germany. If you have any concerns regarding our products, please
contact ProductSafety@springernature.com

Printed and bound by CPI Group (UK) Ltd, Croydon, CR0 4YY

28/04/2026

02098482-0001